Molecules, Microbes, and Meals

Molecules, Microbes, and Meals

THE SURPRISING SCIENCE OF FOOD

Alan Kelly

OXFORD
UNIVERSITY PRESS

Oxford University Press is a department of the University of Oxford. It furthers the University's objective of excellence in research, scholarship, and education by publishing worldwide. Oxford is a registered trade mark of Oxford University Press in the UK and certain other countries.

Published in the United States of America by Oxford University Press
198 Madison Avenue, New York, NY 10016, United States of America.

CIP data is on file at the Library of Congress
ISBN 978–0–19–068769–4

3 5 7 9 8 6 4

Printed by Sheridan Books, Inc., United States of America

For my family, near and far.

With art direction and food photography by Brían French, and microscopy images from colleagues in University College Cork.

{ CONTENTS }

{ ACKNOWLEDGMENTS }

One of the themes of this book has been how every food product or meal has a huge and often underappreciated amount going on behind the scenes to make it what it is. The creation of this book has been just like this, and there are many individuals without whose support it would not have been possible, and whose contributions have ensured that the outcome is as palatable and pleasurable as I hope you will find it to be.

First, I wish to thank Jeremy Lewis, my editor at Oxford University Press, for his initial positive reaction to my proposal, and subsequent support, advice, and patience throughout the writing and editing process. I would also like to thank others at Oxford University Press who have been very supportive in the production of this book, including Anna Langley and Lincy Priya, as well as Victoria Danahy for her expertise and contributions during the copy-editing process.

Next, I owe a huge debt of gratitude to Brían French, a graduate of the university where I have spent my career teaching and studying food, University College Cork (UCC) in Ireland. Brían created the wonderful food photography for this book and supplied inputs and advice on its art direction throughout the process, as well as giving feedback on the entire draft text. I also wish to acknowledge the willingness of Dr. Andre Toulouse of the Electron Microscopy unit at UCC to collaborate on this book project when I first proposed the idea, and the many hours Suzanne Crotty then spent examining a selection of products and samples, from her kitchen and mine, while being constantly open to my increasingly strange requests. Other images and ideas were readily supplied by UCC colleagues, for which I wish to thank Dr. Stefan Horstmann, Paddy O'Reilly, Dr. Jennifer O'Mahony, Dr. Kamil Drapala, Dr. Juliana Villa Costa, and Dr. Seamus O'Mahony for their interest and generosity.

I would also like to thank colleagues and friends who read draft chapters and provided detailed feedback, suggestions, or corrections to my science when I veered outside my areas of expertise. In this light, I particularly thank Dr. Shane Crowley for many interesting discussions and suggestions, as well as a steady supply of both ideas and articles, as well as Dr. Stefan Horstmann, Dr. Eileen O'Neill, and Dr. John Morrissey for their critical reading and inputs in the areas of cereal science, meat science, and microbiology, respectively, as well as Dr. Roisin Burke of the Dublin Institute of Technology for her feedback in the areas of the science of the kitchen and molecular gastronomy.

Finally, I wish to thank my wife Brenda and my children Dylan, Martha, and Thomas (as well as my extended family) for their support throughout the writing of this book and for encouraging me to try doing something different for a while. Taking the time to make a long-held ambition of mine a reality has led to a hugely enjoyable and rewarding new turn to my life and career.

{ 1 }

Introduction

FOOD FROM THE INSIDE OUT

First, I have a confession to make.

I am a food scientist.

I have spent a large part of my life in a white coat, or working with students in white coats, studying, analyzing, and creating food products, subjecting them to a variety of processes and tests to see what happens, and occasionally, very occasionally, even tasting them.

This is my passion, and to me is one of the most exciting types of scientific research in which I could be engaged, where the challenges are complex and really interesting, but in every case relate in some way to something central to everyday life. Food science is probably the only field in which a scientific experiment can lead to a change that can have a measurable impact you can point to on a shelf or plate within a matter of days.

Also, it is great to work in a field of science where sometimes, if your experiment doesn't work, you can at least eat it!

However, I accept that, for many people, this is not food.

Food comes on a plate.

Food is an art.

Food is an experience.

Food is pleasure.

Food is life.

Food is not something to handle with a white coat on, not something to deconstruct in test tubes, and certainly not anything to do with chemicals. Definitely not anything to do with chemicals.

Food is not science; food is art.

People today know what they want from the food they buy. They want a wide variety of safe, natural, convenient, nutritious, great-tasting food.

They likewise know what they do not want. They don't want processed food, they don't want chemicals in their food, they don't want preservatives. This presents those who provide that food with great challenges as, to deliver the things consumers want, they often have to avoid the very tools they have traditionally used to achieve these goals.

In this book, I want to explore the contradictions at the heart of our understanding of food, which arise in part from the fact that food is both science and art.

Food is science; food is art.

Art itself is intrinsically linked to science. The way in which pigments mix and respond to light on a painting, or the way in which sound waves interact with our eardrums in a concert, are all intrinsically scientific considerations. To see them as such does not diminish them, but I believe rather shows their beauty in a new light (or sound, as the case may be).

The great physicist Richard Feynman once referred to how a scientific appreciation of a flower, in terms of the complexity of the processes and structures therein, does not in any way diminish the beauty of the flower: "science knowledge only adds to the excitement, the mystery and the awe of the flower. It only adds. I don't understand how it subtracts." To me, exactly the same argument can be applied to food.

Whether we describe the flavor of a fine wine with the verbose and occasionally preposterous language of the connoisseur or with a tabulated list of the responsible chemical compounds generated by an expensive mass spectrometer, it can still be a thing of wonder and discovery.

I hope to make a case that appreciating food from the perspective of the scientist does not diminish the wonder or mystery of food, but rather can enhance it further, just as Feynman argued for the flower. Whether seen with the eye on a plate in the finest restaurants or magnified millions of times under an electron microscope, as we will see in images in this book, food can take your breath away with unexpected appearances and secrets.

What makes food science one of the most fascinating, in my view at least, fields of science is that it was an art long before it was a science. We have explored complex scientific principles through the ancient method of trial and error for millennia and, as a result, many key scientific concepts and tricks have passed beyond science into common practice and craft. We have practiced fermentation to make bread, beer, and wine for far longer than we knew what was causing this or how to control it. We have fueled great voyages of discovery by preserving food using salt and made food safer and less perishable by heating it or keeping it cold. We (or at least the Incas) even apparently understood how to freeze-dry food in the cold, dry mountain air long before the physical principles involved were ever articulated.

Today, this leads us to a common perception that "old" or familiar science is good and not really science, but "new" science is somehow bad and suspicious. A few years ago, a large US yogurt manufacturer made a marketing point out of their rejection of science in their product. Nonetheless, we will see in Chapter 7 how yogurt wouldn't exist if it wasn't for highly complex microbiology involved in fermenting the milk, plus a dash of advanced protein chemistry to explain how the milk proteins form the creamy gel they ascribe to mysterious nature. Perhaps science is okay as the behind-the-scenes stuff you don't want to acknowledge as giving you the actual product in question, once you don't contaminate your packaging (or consumers' minds) with its pesky weird terms and explanations that ruin all the mystery. Science shouldn't be the star, but rather is like some unglamorous grimy backstage crew who never get invited to the award shows.

Likewise, we instinctively eschew the idea of chemical preservatives, but accept our smoked salmon, pickled fruit, sugar-laden jam, salted butter, and alcoholic beverages without hesitation, and without acknowledging or even recognizing that the defining ingredients of all those products were first used or accepted specifically because they are chemical preservatives, again long before the term was ever dreamed up.

Many food products are triumphs of science. For example, of the dairy products, cheese, with its teeming ecosystem of microorganisms, enzymes, and reactions, is like a case study of many different fields of biology. a synergy of molecules and microbes. The much-beloved properties of ice cream, on the other hand, are essentially entirely the result of physics, with a little help from chemistry.

By trial and error, mankind has practiced food science since the dawn of civilization. Without a basic understanding of food processing and preservation, all people would have to live where food is immediately available for consumption, and cities would never have emerged. Without the ability to process and preserve food, our supermarkets today would be far smaller premises, their shelves absent of the food that is available to us only because scientists worked out how to make them last long enough to transport over long distances.[1] Of course, I do not deny that science sometimes goes too far, and not all developments have been embraced by consumers or have been overall positive, but the overwhelming direction of progress has been positive.

At certain times, specific imperatives have driven bursts of development in food science. For example, when asked which 19th-century Frenchman contributed greatly to the advancement of the preservation of food by heat, many people will think of Pasteur and his prepared mind. However, decades earlier, Napoleon had catalyzed the development of what would become canning of food by the pragmatic step of offering large sums of money to the finest brains in France. The challenge was for them to figure out how to make food his armies could take with them on their hostile interventions into countries where it was reasonably assumed the local population would be unlikely to invite them in for dinner on their marches.

Similarly, much more recently, NASA became one of the greatest sources of innovation in packaging, formulating, and processing food with the highest possible levels of safety. This was due to the clear need to keep astronauts happy, fed, and optimally nourished while using the universe's simplest kitchens and larders, without the slightest risk of getting a rumbly tummy while tethered in a space suit far from home. This has led to benefits like a system for managing food safety,

---

[1] Illustrated beautifully in a video called "A World Without Food Science," by an American professional organization called the Institute of Food Technologists, which can be found easily on YouTube.

called HACCP (which stands for hazard analysis and critical control points), which is applied almost universally today, as well as developments in freeze-drying, tailoring nutritional profiles of food, and the production of bakery products that are less likely to produce instrument-endangering crumbs.

As food science has progressed, the craft by which products like beer and bread and cheese have been made and instinctively understood for centuries has been gradually complemented by a deep scientific understanding of the processes involved and how to control them so as to consistently get the best and safest products, just as consumers demand them.

Nonetheless, a word people don't want to hear about in connection with food is "chemicals." A few years ago, a European survey reported that 83% of consumers were worried that their food and drinks contain chemical substances. Of course, food is composed of chemicals! Some of these are called proteins, and sugars, and vitamins, and more—these are all chemicals! Food scientists might indeed add chemicals to your food or just change and rearrange them. Every food product is nothing more than a collection of chemicals, some simple and some incredibly complex, transformed through reactions induced by processes such as cooking, that interact and react with us before and during our eating.

When faced with a choice between natural and artificial flavors in something, there often seems to be no contest. One comes from nature, and the other is "chemicals." Imagine if you had to state on a label that a product contained compounds like butanoic acid, hexanoic acid, 2,4,6-heptatrione, 5-hexyl-dihydro-3H-furan-2-one, 2,5-dimethyl-4-hydroxy-2H-furan-3-one, and 2,5-dimethyl-4-hydroxy-2H-furan-3-one-glucoside. How would that look? Answer: chemical, artificial, nasty! A paper I just checked in a journal called *Food Chemistry*, however, listed those as the major flavor compounds in fresh strawberries, and each one can be linked to a particular component of the flavor we associate with these lovely fruits, and all are perfectly innocuous despite their names.

The successful production of every dish and food product, whether in a kitchen, a small farmhouse cheese plant, or the largest brewery, requires the application and control of both science and technology, explicitly or implicitly. Here, and throughout the book, I use the terms "science" and "technology" with the meaning that the former is about understanding, while the latter is about making and creating. In the context of food, technology includes every operation we apply to the raw materials of food before we consume them, from simple steps like washing an apple or refrigerating sushi to the complex multistage processes involved in producing a beautiful dish in a restaurant[2] or a product like powdered infant formula. Understanding, through science, how these processes affect the chemistry and biology of food and the unseen microbial hitchhikers they may contain will be explored in detail in the pages to come, as well as how they ultimately affect our senses when eating, from crunch to sniff.

---

[2] By this definition, every chef is a food technologist!

The complementing of the art and craft of food with science has perhaps been slowest to develop in one domain where our relationship with food remains very approximate, instinctive, and dependent on the person responsible, and that is the kitchen. Many of the transformations between raw materials and ingredients and fork remain mysterious, although unequivocally scientific phenomena. The worlds of the scientist and the chef, both intimately concerned with this subject called food, remained for a very long time remote and disconnected, even being rarely taught in the same educational institutions. Practitioners of these twinned disciplines, separated at birth, often couldn't even talk to each other, using as they did such different languages and perspectives on the same topics.

However, even this is changing today, with food scientists increasingly working in kitchens, food companies hiring chefs to prepare the ideal products that consumers will want and value, for scientists to emulate and produce, and chefs increasingly viewing food as a scientific topic. Some of the world's top restaurants for the last decade have kitchens that more closely resemble laboratories, and the emergent idea of molecular gastronomy has arguably reunited these estranged twins. Even beyond the science fiction kitchens of such establishments, far beyond the geographical or financial reach of the vast majority of consumers, every person who prepares and cooks food in a kitchen is engaging in a highly scientific set of activities and unleashing principles of biology, physics, and chemistry that he or she does not apply as casually in any other part of life, besides the everyday miracle of living itself.

Every kitchen is a laboratory, and every meal is an experiment; it just depends on how you look at it.

When trying to persuade school pupils to consider studying food science at university, I always come back to one cliché; there is more to food than meets the eye. We take food for granted, at least those of us who are lucky enough to live in a society where safe, affordable food is readily available (and who often forget the large proportion of the world's population for whom this is not the case).

Food isn't complicated, or difficult, or challenging, it is just food.

The average consumer picks up a bag of salad in a supermarket and sees something plain, simple, and healthy. The food scientist picks up the same bag, understands the complex materials science that led to the creation of the bag itself (with precisely controlled abilities to control the access and impact of gas, moisture, light, and heat, and exactly the right mechanical strength for the handling it will get throughout its life cycle), the gas within (not air, but a carefully selected mixture of inert gases to keep the contents fresh), and the plant-derived complex changing biological entity within, the constituent molecules of which provide color, crisp texture, flavor, and nutrition. Same product, different viewpoint.

An understanding of food science and the processes we apply to it before consumption does not diminish the wonder of food, but rather enhances it, and in this book I hope to persuade you of this fact.

The topics cover both industrial food processing and production and small-scale processing in kitchens, restaurants, and homes, and a goal is to demonstrate how the principles at both scales are basically the same. This book is neither an attack on nor a deliberate defense of the practices, politics, or ethics of the food industry, with the focus being on the tools used and their principles and effects, not the users and what they choose to use them for.

Another key aspect of food is of course its nutritional significance and how the components therein, from vitamins and minerals to amino acids and sugars, affect our bodies and our own biology. I consider these aspects in terms of where the molecules concerned come from and what happens to them during the journey to our mouths, but, at the moment of swallowing, food moves into a different realm of science, and one that I do not dwell on in this book, as it is already the basis of many other books, and beyond my own expertise to discuss.

To finish, I started this section with one confession, so I end with another.

Not only am I a food scientist, I am not a foodie.

I cannot cook beyond the most basic of dishes.

I have a ludicrously narrow set of food preferences for eating, including an inappropriate love of junk food.

I have never shared a picture of a dish on social media.

This is not the prism through which I have written this book.

Nonetheless, I love food very deeply, from the bottom up as it were, from molecules to meals. We usually approach food from the outside in, as we open a package, peel back an outer layer, or bite through something, but my viewpoint is rather from the inside out, as molecules assemble, interact, and transform to give food the character we expect.

In this book, we explore the inner beauty of food and see how that ultimately leads to our inner beauty, and so we look at food from the inside out. This principle underpins the choice of illustrations for this book, which include a combination of photos of food as we might be familiar with, albeit often with a twist, and photos of food as the reader might never have seen it before, mostly taken under powerful microscopes to show the life (including microbes), structures, and interactions going on in even the most familiar products. One of the most powerful tools for understanding the properties and nature of food is an electron microscope, with which images of much finer resolution and greater magnification than those for a light microscope can be obtained, and most of the images shown come from such a microscope in my university.

Through this approach, in the chapters to follow, I hope that readers will come to consider food in a new way and will see a whole new set of wonders in every dish or package. We will look at food from both sides now, up and down, and still somehow we might see that we didn't really know food at all, until all these angles are considered.[3]

First, we consider where food comes from, and then look in subsequent chapters at the key constituents of food, specifically proteins, fats, and carbohydrates (as well

---

[3] With apologies to Joni Mitchell.

as water). The many ways in which microorganisms (such as bacteria and fungi) are significant (for positive or negative reasons) in food are then explored, leading on to how processing seeks to make food safer and more stable, whether through heating, drying, cooling, or many other approaches (including packaging). It will be seen, as indicated in the title of this book, that microbes such as these are actually a key part of the science of food, for better or for worse. Many of the examples I use are related to dairy products, partly because this is my own area of greatest expertise, but also because seeing how this highly complex system works can be used as a case study from which to explore many different phenomena in all kinds of other food systems.

In the final chapters, having moved from raw materials and food components through to how products are produced, we look at what happens when we eat, how products and innovative dishes are created, and then we explore the relationship between what happens in a processing plant and what happens in a kitchen, and consider the science and technology behind every meal and dish.

In this way, my goal is to take the reader on a pathway that leads from molecules to meals, and uncover and indeed celebrate the science behind every mouthful we consume.

On the Origin of the Spices
(and Other Foodstuffs)

The origins of food

The beginning of the story of food is what is termed food production. This might sound logically like the process of making food, such as a chef or food company might, but this term is rather generally used in food science to refer to the so-called primary production of food, from growth of crops to harvesting of fish and minding and milking of cows. Primary production is, for example, what farmers do, producing the food that is brought to the farm-gate, from where the processors take over. So the food chain runs, according to your preference for a snappy soundbite, from grass to glass (for milk), farm to fork, slurry to curry, or (taking the food chain to its logical conclusion, and including the role of the human gut charmingly but appropriately in the chain) from farm to flush.

But where do these raw materials that are yielded by primary production actually come from?

It is often said that all things found on earth can be divided into categories of animal, vegetable, and mineral.[1] To these could perhaps be added two more categories, microbial and synthetic (man-made). Within these five groups can essentially be placed everything we know as food, so using this classification to consider where our food comes from seems like a good starting point for this book.

Perhaps the simplest group to start with is minerals, which might intuitively seem an unlikely source of foodstuffs (do we eat metal or rock?), until we consider where salt comes from and how much of it we add to our food (in other words, probably too much).[2] Our bodies, however, absolutely need for us to consume certain metals and other chemical elements to survive, beyond the sodium and chloride we get from salt, and so many extracted minerals find their way from deposits in the earth

---

[1] A distinction made most popular through a TV show in the 1950s, in which experts were challenged to put unfamiliar objects into one of these categories.

[2] Confusingly, in Ireland, the term "mineral" is widely used as a generic term for a product type that rarely contains such compounds, which includes carbonated soft drinks, such as Coca-Cola and 7-Up. The origin of such confusion is unclear!

into food products. This is particularly important where their biological effects are a desirable outcome (such as in carefully formulated nutritional products). In addition, products such as milk contain minerals like calcium, magnesium, zinc, and more, because the infant or calf needs them to thrive. Interestingly, though, as we will see later, the tiny quantities of such minerals present also have pretty dramatic significance for the properties of food, such as making milk white and influencing the texture of meat.

Moving on to another category of "things" that are often found in food, we come to the vast and teeming world of the microorganisms that thrive invisibly. We will see in later chapters of this book how much of food is dependent on the actions of these agents, from the bad (food poisoning) through the ugly (food spoilage) to the good (from fermentation of beer to the production of valuable food ingredients by bacteria).

In terms of man-made materials, entirely synthetic foodstuffs are perhaps rarest of all, but examples might include artificial sweeteners designed by chemists to confer the desirable sensation of sweetness without the undesirable accompanying calories. We might also include here starches that have been chemically modified to manipulate the way in which they confer texture and structure on food products.

We now turn to the most conventionally alive sources of food, which together represent the origin of the vast majority of food we consume: vegetables (or plants more generally) and animals. Our food largely consists either of the whole organism or something the organism has produced that we have extracted or recovered while the organism continues on, like apples from a tree or milk from a cow.

Every living thing, from chickpea to chicken, is a complex biological system, which remains alive through a bewildering array of metabolic reactions and forces, from electrical impulses to chemical reactions. These cumulatively result in phenomena such as movement, growth, conscious thought, and the writing of books about food.

All life on earth, as far as we understand it, is built around the properties of one element, carbon, and most of the molecules that are critical to animal and plant life depend on this element above all others. We refer to molecules that contain carbon as being organic and those that do not, such as the salt previously mentioned, as inorganic. In a sense, carbon can be regarded as the spark of life, without which life, at least on earth, would not be as we know it, to paraphrase a famous non-earthbound engineer.

Interestingly, while many food products are today available in organic and non-organic versions, depending on the method of production and avoidance of certain agricultural chemicals, a scientist could use these terms in very different ways to distinguish food products. All food products containing carbon could be regarded as organic; the nonorganic shelf in a supermarket categorized in this way would be rather bare, containing mainly pure water and salt.

Plants and animals clearly differ in certain obvious characteristics, such as movement, responsiveness, and tendencies toward robust communication, but at a fundamental chemical and biological level they are not really all that different after all. Take a potato and a person and analyze their chemical make-up and the types of molecules they contain, and most will be found to be in common.

Potato and person, not to mention penguin and pineapple, are all built around the same fundamental unit of life, the cell. Cells are the basic structural unit of life, a microscopic blob of gooey liquid, surrounded by a protective membrane, and in some cases (plant cells, for example, but not animal cells) a more rigid wall, within which sits the nucleus. This nucleus holds the DNA (a much more familiar term than its full name, deoxyribonucleic acid) that is the masterplan for the biological activities in the cell, and beyond that the organism as a whole.[3]

Cells we find in our brains may differ from those in our stomach from those in our blood, but the same basic functions (and DNA masterplan) exist in all. Across different forms of life, the details of the cell's architecture might vary (and in some cases, like viruses, differ pretty wildly), but most of the reactions those cells engage in on an ongoing basis will be similar, all depending on the same principles of genetics and wondrous properties of DNA for the propagation of their species.

Any comparative analysis across the broadest scale of visible living things would also find certain classes of molecules to appear very commonly, such as proteins, carbohydrates (sugars), and lipids (fats or oils). These, all carbon-based molecules, are the basis of every food product and, as we will explore throughout this book, are the sources of every aspect of "foodiness" we attribute to food products, from their appearance to their texture and flavor and the changes they undergo when we cook or eat them.

Other molecules present in these living things that are important in our food, even if present at tiny levels, include the inorganic minerals mentioned earlier and vitamins. We tend to associate these with nutritional benefit, because of our bodies' need for them on an ongoing basis to sustain life, but they can have quite significant effects on food properties too. For example, minerals can hugely affect how proteins act and give texture to food, while many vitamins help to preserve food through preventing undesirable chemical reactions such as oxidation, which is the degradation of certain kinds of lipids into frequently awful-smelling and -tasting products through reaction with oxygen.

From molecules to meals

Every single meal and every food product can be described by a list of raw materials and ingredients, plus a set of instructions as to the steps that we will apply to these. Every food label includes a list of what is in the contained product, while every recipe in a cookbook starts with a list of things to be assembled in order to create that meal. Every kitchen contains a basic set of raw materials, whether vegetables,

---

[3] We don't think of DNA as a food component, but much of what we consume was once alive and, when alive, depended on DNA for its very aliveness. Nonetheless, DNA is readily broken down in our bodies, making sure that "you are what you eat" remains a figure of speech.

types of meat, spices, salt, and so forth, and the art of the chef (ridiculously simplistically put) is to assemble different combinations of these in different proportions to get a range of different and desirable outcomes.

In either case though, we could imagine erasing the list and replacing it with a new one, which describes exactly the same mixture of materials present in the end product. However, this list might look a little less homely, not containing familiar terms like carrot, coriander, and chicken, but instead perhaps carotene, actinomyosin, casein, ascorbic acid, amylopectin, linoleic acid, and a multitude more. These are the true raw materials of food, the building blocks that comprise the individual ingredients, now expressed in terms of their chemical essence. When found together, naturally or as we mix them, and subjected to the chef's or producer's art, they lead us to the end result we call food. Each molecule has an identifiable role, with the smaller molecules perhaps conferring flavor and the larger ones texture, sometimes individually and sometimes through their interactions. Many of these change or transform as we process or cook the mix, but the end result is the same.

So, all food is made of molecules, meals are made from molecules, and molecules come from many places, but most commonly from nature itself. How we go from molecules to meals is the magic of food science.

The transformation to food

Most food begins as something that is alive or is produced by something that is alive, but that has (usually) ceased to be at the point at which we consume it.[4] We can trace the desirable properties of many food constituents back to their function in their place of origin. The texture of steak reflects the muscular movements of the cow that it comes from, as surely as the soft white texture of fish can be directly traced to the act of swimming and the starch in so many cereal-based products was intended by nature to store energy absorbed from sunlight.

Of course, most of these molecules developed through millennia of evolution but not for the purpose to which mankind has learned to put them. Muscle is not designed to be cooked, while cows' milk was intended for calves, not humans, and was certainly never expected to be heated to the temperatures it routinely encounters while being processed into dairy products.

It is in fact remarkable that so much of nature around us turns out to be suitable for, and indeed pleasurable for, human consumption, particularly after it has been subjected to processes nature never envisaged. Of course, nature is full of consumption, whether of antelope by lion or grass by cows, but what is unique about

---

[4] Common exceptions to this include yogurt, which at point of consumption is generally very much alive indeed, in terms of the bacteria within that have caused the fermentation that gave us the product, now off duty but still present. In addition, biological processes will continue in fruits and vegetables after harvest, which can change and ultimately impair their quality.

humans is the way in which we have learned to transform raw materials into a different state before we eat them.

Many of the goals of food processing as it has developed throughout history have concerned taking promising but not quite perfect raw materials and transforming them into a form that we find easier and more enjoyable to consume. A further key driver in the history of food processing has been the unavoidable fact that, just as these materials provide ideal nourishment for us, they are also undeniably attractive to many other forms of life, most of them invisible to us but still capable of inhabiting or changing the food in ways that may be alternatively interesting, unpleasant, or downright lethal.

Before we consider in the chapters to come the science of each of the types of molecules that are almost uniform to all types of food and the ways in which these are manipulated to give us our food, whether in a package or from a kitchen, it is worth considering how we came through history to where we are today in terms of our relationship with nature, as concerns our dependence on large parts of it for our daily sustenance.

It has been reasonably argued that the development of civilization itself owes much to the discovery of basic principles of food processing such as heating and refrigeration, as well as chemical preservation of food. Understanding of how such phenomena made food last longer and less likely to induce illness allowed food to be transported over greater distances, allowing early civilizations to develop cities, in which consumers of food were geographically separated from its producers, which freed up the consumers' time to concentrate on other things, like becoming later civilizations.

In later centuries, the ability to transport food over longer distances, requiring stability over much longer periods, facilitated exploration, including the great voyages that discovered America, explored the poles, and mapped the rest of the globe in great precision. Pickled and salted barrels of food provided critical, if somewhat unappetizing, sustenance for such travels. Then, as history progressed, the definition of the target distance over which food needed to be transported stretched, from the distance between farm and city to the distance between countries (at peace for export or at war for conquest), and even the distance from the earth to the moon.

Going much further back, a theory called the Cooking Hypothesis[5] purports that a critical step in the evolution of humans was the discovery of the use of fire and other sources of heat to process food, converting it into a form from which nutrition can be extracted with far less effort and time than if we consumed raw meat or vegetables. Cooking effectively partly externalizes the digestion of food, starting its breakdown and conversion into more readily absorbed forms outside the body, and meaning that humans need to spend significantly less time eating, chewing, and digesting their food than any other species, none of which developed the habit of

---

[5] Articulated in a book entitled *Catching Fire: How Cooking Made Us Human* (2009) by Richard Wrangham, a British scientist at Harvard University.

cooking their food before consumption. Again, such saved time freed up the energy and time of early humans, and this is thought to be a key step in our development, as it provided energy for accelerated brain development.

The variability of raw materials

Because many raw food materials have a biological origin, a wide variability in their composition and properties can apply, which adds to the challenges of producing food of consistent and predictable quality. Just as all chefs will closely scrutinize candidate ingredients for their dishes and reject the undesirable ones, or modify their recipes based on a range of often intuitive criteria, so food processors have to take into account the fact that the material coming in this week might behave quite differently to that which came in last week.

For example, the processing of milk into dairy products is complicated by the fact that the composition of raw milk can vary widely for a range of reasons relating to what cows do and eat. There is a clear relationship between the diet of cows and the levels of components such as fat and protein in the milk they produce and the exact individual fat and protein components present. This can be used to deliberately alter the type of milk produced, but also means that at different times of the year the milk coming into a factory is quite different to that coming at other times, simply because of the amount of grass consumed by the cows. A cow could be regarded as a (very efficient) machine for turning grass into milk and, in some countries, like Ireland, grass is indeed the main feed.[6] As a result, "summer milk" can be very different in its properties to "winter milk," particularly in terms of the types of molecules called fatty acids that are present, which have major implications for the production of products such as butter.

So, as this example shows, the composition of raw materials to be processed into finished food products is not fixed and not always easily predicted, and so a number of processes have been developed to alter or standardize the composition of raw materials, some of which will be explored in later chapters.

The same factors of seasonal or other variations of course apply to the availability and maturity of fruits and vegetables, meat, and cereals, and indeed working around nature's own rhythms and cycles defines much of what we see on market shelves and menus. For all food products, the best way to start off with a safe and high-quality end product is to start off with a safe and high-quality raw material, and surveillance and quality assurance must start right at the very beginning of the food chain.

---

[6] To take advantage of this, most cows in Ireland have calves in the spring, so that they are producing milk at the time of maximum availability of grass. This is called synchronized calving, which is not to be confused with synchronized swimming, as it does not involve large numbers of cows in a pool performing in a graceful manner.

The very availability of food is also hugely dependent on much larger factors than we can easily control, as when drought results in the failure of crops and the denial of available food for entire populations, leading to widespread death, migration, or war, as seen throughout history in examples from the Irish potato famine of the 19th century to famines in Asia and Africa across the 20th century and today. It can never be forgotten that the ready access to high-quality and safe food at affordable prices is a privilege not shared by huge sections of the world.

At present, there is significant debate and discussion around expanding our classification of those parts of nature that we regard as suitable for consumption to include a massive (by weight all around us) source of nutrients, especially protein, which is insects. In some countries, insects (especially beetles, but also caterpillars, bees, locusts, and crickets) have been consumed without controversy for millennia, while today insect-based food products or products containing powdered extract from insects (and so less visibly anthropoid[7]), are appearing in many others. Using insects as a raw material for food has many potential advantages, including their having high levels of proteins, vitamins, and minerals by weight, being more environmentally friendly as a food source than meat in terms of amount of plant material consumed to make a certain weight of protein, and their "farming" requiring far less space and water than other types of food production. However, while adoption is rising, in some countries it has plateaued or even decreased, and consumer resistance to the idea of eating "bugs" has been too big a barrier (so far) to widespread adoption in Western countries.

From nature to nourishment: Plants, fruits and vegetables

Our usage of the natural world for food has become an exceptionally efficient process. We have divided all the earth provides, mobile or not, into (A) that which we can eat and (B) that which we cannot eat.

If we consider the plant kingdom, for example, all that which is (mostly) green and grows on the earth has been triaged into: edible (in part, in whole, or producing or containing parts that are edible); not edible but functional for aesthetic pleasure (many flowers); not edible but useful for other purposes such as supply of materials, oxygen, shelter, or aesthetic value (trees); and not edible but insufficiently useful or pleasant to warrant tolerance (most weeds).

Within the plant kingdom, we call those products that we consume in whole or in part vegetables or fruit. Look at any biology textbook, and the structure of a plant can be found depicted in anatomical precision, from roots through shoots

---

[7] The eating of insects is named using the wonderfully mysterious sounding term entomophagy, not to be confused with anthropophagy (cannibalism). The cultivation of insects for food is sometimes called minilivestock (I love that!).

and stems to leaves and flowers, all with specific functions, from storage to support to reproduction. A botanist and a food scientist will look at such a diagram with different perspectives, though, as the latter will look in terms of which bits can be eaten, either in whole or in part, safety, desirability, nutritional value, and whether the plant will survive such intervention.

Sometimes we choose to eat the roots of plants that grow under the ground, as in the case of carrots and turnips, while in some cases we are interested in the leaves (lettuce, cabbage), stems (rhubarb above ground and the subterranean potato) or flowers (broccoli, tea, and cauliflower, none of which are commonly admired in a vase for their intrinsic beauty). Of course, we get cereals from grains such as rice, wheat, maize, and barley, while beans and peas are what we call pulses and are actually plant seeds (as are coffee beans and nuts).

For other plants, our interest is in their fruits, and we return to the same source repeatedly to harvest these products, as with apples, oranges, pears, grapes, and many more, while some types of plant are of value because they yield oils, including rapeseed and sunflowers. Some fruits contain seeds through their mass, and these we call berries, including blackberries, raspberries, and grapes. Strawberries, interestingly, are a true curiosity of nature, being technically neither fruit nor berry, as what might be pictured as seeds on the surface are actually the fruit part of that plant while the juicy main part is the base of the flower.

Finally, herbs and spices, like coriander, chili, clove, cumin, curry, and cinnamon (to take but one letter from the immense spice rack of nature), come from plants too, being dried leaves (as in the case of herbs) or, in the case of spices, roots, buds, fruits, or seeds that have particular and long-valued aromatic characteristics.

Plants are very efficient biological machines for turning what nature provides into food, absorbing sunlight and carbon dioxide from the air and nutrients from the soil and storing the resulting nutritious outputs in a form that makes them easily digestible to humans.[8] These processes are biological and dynamic, which makes obtaining the optimum products a time-sensitive business, depending on season and time at which the parts of interest are harvested, before biology has moved on, and what results is of less value to us.

Plants also have sophisticated mechanisms for procreation and replication, which are fundamentally different to those of animals, thanks once again to their, by definition, stationary existences. For example, fruits and flowers sometimes get help to spread their seeds by taking advantage of the attention of other organisms, as in the case of bees spreading pollen, or the wind in dispersing light seeds. Animals that eat fruit also spill seeds from sloppy fruits that contain lots of them, or even sometimes excrete undigested seeds in their feces that, if they land on promising soil, become a new place for growth of a progeny plant.[9]

---

[8] The key to this miraculous ability is the presence of a compound called chlorophyll, which gives the characteristic green color to most plant life through a process called photosynthesis.

[9] Which is, let's face it, a pretty weird strategy for achieving combined goals of reproduction and house move.

Fruits also change through ripening, thanks to reactions converting starch to sugar, and indeed are most readily differentiated from vegetables by their sweetness, often combined with a touch of acidity, and the action of enzymes breaking down cell walls (all phenomena to be explored in more detail in coming chapters). Fruits are the only parts of plants intended to be eaten, thanks to their role in ensuring the transfer of seeds and creation of new plants, and so it is not surprising that their flavors are typically deemed more pleasant to us and match better what we expect from food than those parts of plants (vegetables) not intended for such a fate.

Go too far with these processes, though, and the result is less appealing. Bananas continue to ripen as well when removed from their tree as they do still attached; as starch is systematically deconstructed to simple sweeter sugars, the pectin that gives cell walls their structure is gradually broken down to a creamier, mushier texture, and enzymatic reactions in the skin act on compounds called phenols to give black discoloration.

A banana, like many plants (and indeed more broadly many food products), thus goes on a journey, the metaphor so beloved of writers and filmmakers, from immature and inauspicious beginnings through to maturity and a blossoming appeal, in which prime they can be found at their best, before a slow (or sometimes not too slow) decline in later age. As we will see, a large part of the goal of food processing, from the field to the fork, is in controlling the rate at which these changes take place, to maximize the golden period in between immaturity and overripeness in which we can best admire nature's fine work.

The texture of plant-derived foods, as for so many food products, can be related directly to biological structures. We find firm structures in parts of a plant expected to stand upright or bear their own weight and that in food confer strength and crispness, and different and less-structured textures we find in parts of vegetables like potatoes that do not have to withstand the forces of gravity during their time in the earth. Plant cells are very different in their structures from animal cells, and a key difference is their rigidity, which is due to the lack of need for the flexible structures required by forms of life for which movement is a key activity, and the greater requirement for strong and stable governance of their shape. Plant cells use for this purpose, in particular, molecules called cellulose and lignin, which are related to starch but are characterized in particular by their strength and difficulty of digestion.[10]

Leaves and stems are shaped to maximize the surface area through which they can absorb energy from sunlight, while being relatively full of air cells that make them susceptible to damage like bruising and shrinkage during cooking. Similarly, ripe strawberries are also heavily laden with air, and once we physically damage them this is released and crisp structures turn mushy.

---

[10] When selecting materials for building a house, we want to use those that are least likely to break down or weaken in any environmental conditions to which they might be exposed. The same principle underpins the presence of cellulose in plants.

Plants and flowers are also valued for their color of course and, when considering the color of plants, we think of course of green (forty shades, or more, thanks to chlorophyll), but also sometimes yellows and oranges (because of the presence of compounds called carotenoids, also found in milk thanks to cows absorbing them from plants), and sometimes purples and blues (because of compounds called anthocyanins, which are interesting pigments in the kitchen thanks to their behavior under different conditions of acidity or alkalinity). Unduly harsh treatment can cause molecules to mix that nature did not intend, leading to brown or black colors.

One of the most common traditional uses of a plant-derived food product is probably tea, which has been established as a beverage all around the world for millennia, probably originating in China. All tea comes from a single kind of plant, with the Latin name *Cammelia senensis,* but, as we will see so many times in this book, tiny differences at various points from bush to cup lead to massive differences in the final product (a form of chaos tea-ry). For example, green and black teas differ in how the leaves are processed after harvest, while, as is the case for wine, differences in where the *Cammelia* bush is planted and local climatic conditions, as well as when it is harvested and how, all result in tiny biological differences in relative levels of certain compounds and chemical reactions that ultimately lead to very different characteristics of the final beverage. The conversion of tea plant to tea drink involves more biological processes, in this case the encouragement of the activity of enzymes in the leaves that produce flavored, colored compounds (tannins) that give tea much of its most desirable attributes.

Peer down a microscope at any plant-derived food and the botanical origins are immediately clear. For example, Figure 2.2 shows several views of the internal structure of a banana (the original of which is seen in Figure 2.1), none of which would immediately convey their origins.

This is the first example in this book of an image taken with an electron microscope to gain a new perspective on the structure, origins, and transformations of food, to help us see familiar foods from the inside-out, so it is worth here pausing to explain briefly how such an image is taken. When we examine something using our eyes, or a traditional microscope, we are using light, but the resolution, in terms of the size of objects we can see, or the detail we can distinguish, is strictly limited by both the physical properties of light and those of our eyes. Using light and a powerful conventional microscope, we can perhaps see structures or objects as small as 0.2 micrometers (microns) or a fraction of one-thousandth of a millimeter. We can see bacterial cells and blood cells fine at this scale (as we will see in Chapter 7), as well as larger food structures, like fat droplets in an emulsion. However, most of the fine structures within food, such as those created by proteins and carbohydrates, are on a scale much smaller than this, and so cannot be satisfactorily visualized by light.

In an electron microscope, we replace light with a beam of electrons, which allow resolutions of much finer structures than light would allow and can actually magnify objects up to 200 times more than a light microscope. In the main technique used to generate the images used in this book (scanning electron microscope) the food sample was dried, coated in a fine layer of metal atoms, and then placed in a chamber in which a fine beam of electrons was scanned over its surface, which resulted in their bouncing off the food in a manner that was dependent on the structure of the surface being scanned.[11] A detector then measured the properties of these emitted electrons (such as the direction in which they travel), and a computer generated an image based on what shape the object being scanned must have had to have given a pattern of reflections such as that detected. Such micrographs reveal a world of detail and structure in our food that is impossible to see by other methods and have yielded many of the images in this book, such as Figure 2.2.

In this set of images, the clear tubular structures, pores, supporting pillars, and channels of a plant are obvious, particularly in the stem, while the outer surface under high magnification displays a mossy convoluted surface, not immediately relatable to the leathery skin we picture peeling back before we eat.

Figures 2.3 and 2.4 show us the inner and outer structures of a pepper and a tomato, respectively. The tomato shows a highly porous inner structure, scaffolded by folds of inner flesh that have collapsed somewhat in the preparation for the photo. This is surrounded by a porous outer skin, in which the microscopic openings can be readily seen. The interior contains and entraps large amounts of water (thanks in part to the presence of pectin, a water-hungry polysaccharide we will discuss more in Chapter 6), resulting in the juiciness we associate with a ripe tomato. Despite some arguments to the contrary, a tomato is a fruit (technically, it is actually a berry), as it is produced within a flower and contains the seeds to produce progeny. The bright red color arises from a pigment called lycopene, which gradually overcomes the green color of the immature unripe tomato, due to chlorophyll. Following the natural cycles of many fruiting plants, tomatoes ripen when temperatures drop, and sugar levels rise, the fruit becomes less acidic, and the signal for once turns from green to red to show it is time to "go and eat."[12]

The pepper (capsicum) shown in Figure 2.3 is actually biologically closely related to the tomato (as is, unexpectedly, the potato!), and so shows a similar microscopic appearance, with folds of cellular material providing a skeleton around cavities of entrapped water, although with (to my eye) a Harry Potter-esque Dementor vibe, perhaps related to their infernal spiciness. A specific internal membrane of a pepper

---

[11] For those interested in the details, it was a JOEL JSM 5510 Scanning Electron Microscope in the Biosciences Imaging Centre in the Department of Anatomy and Neuroscience at University College Cork, and samples were typically freeze-dried and then sputter-coated with an ultrathin layer of gold and palladium. Images shown were taken at magnifications from 50X to 5000X, and scale bars are not included for simplicity and aesthetic reasons.

[12] Ripening is actually contagious, in that fruits (even different types) release a gas called ethylene when they ripen, which can induce other fruit to ripen. Farmers or processors can control the state or ripeness by borrowing this trick from nature and treating immature fruit after harvest with ethylene to ripen them to demand.

called the placenta surrounds the seeds within the structure, and it is here that a compound called capsaicin is most concentrated. As a pepper matures, like a tomato it changes from green to red (or orange, purple, or any of a color-card of hues depending on specific type) and its level of capsaicin increases, and it is this molecule that gives peppers their intrinsic hotness, through binding to receptors (found all over our bodies!) that register heat and pain and triggering the same sensation we encounter when suffering an actual burn.

Banana, tomato, pepper—all derived from different part of plants, all edible, and all depending for their structure and flavor on the biological processes that characterize all plant life.

Function, form, and food

In comparing the giant kingdoms of life that are plants and animals, one huge difference is that one (plants) makes food from the simplest resources available, without recourse to consumption of other forms of life, while the other does not have that ability to live without ending the life of another and must consume plants or other animals to survive.

To scientifically describe this, we say that plants are "autotrophic," which means that they get what they need from the air, sunlight, and the soil, while animals are called "heterotrophic" because of their need to consume other forms of nutrition. Most plants (leaving aside the occasional carnivorous insect-eating ones) absorb what they need from two main directions, above (the sun and the air) and below (the soil). They have evolved leaves for one purpose, roots for the other, and stems and shoots to join these together, provide a transport system around the whole, and keep the leaves pointing in the right direction; structures like fruits then store the outputs of these activities. Air is everywhere, as is sunlight (to varying degrees) in most places in the world, and, once a suitable patch of soil is present, these will generally provide what is needed throughout a plant's life cycle. So, once a plant finds a spot that supplies most of these basic needs, it is relatively set and doesn't need to move around to get what it requires for life, again a notable difference from animals.

Of course, one irony of nature is that if plants could move, they could extend their life by being less immediately accessible to other hungry forms of life;[13] on the one hand, imagine the sight of a bear chasing a fruity bush across a field or, on the other hand, the scene if all humans needed to do to be nourished was to stay in exactly one spot, regularly absorbing sunlight and air, while their toes wriggled in fresh dirt![14]

---

[13] Plants have had to evolve alternative defensive mechanisms, such as hard barks and thorns, as well as production of bitter and toxic compounds, to try and ward off unwanted attention.

[14] Staying in one place and not venturing out thanks to constant deliveries of takeaway food does not constitute a model for plant-like existence.

While we are down in the dirt, examining the roots of plants and their emergent stems, we might notice other forms of life, simpler yet and similarly unadventurous in their lifestyle, which are fungi like mushrooms. These are biologically related to the fungi and molds we appreciate on the fuzzy surface of Camembert cheese or in veins through a piquant Stilton but reject when they (often the same species) appear on an overlooked piece of fruit. Mushrooms represent perhaps the peak of evolution of this particular branch of the tree of life, achieving a macroscopic structure rather than a microscopic fuzz. Mushrooms, unlike plants, cannot photosynthesize, and so must absorb nutrients from those which have done the digesting for them, for example by coexisting with the roots of trees.[15] The form we eat is the equivalent of a fruit, in that it represents a structure created with the express form of spreading seeds. Structurally, they also differ from plants in that they don't use cellulose as their main brick, but rather something called chitin, also commonly used by insects.

So, going back to the categorization of plants as constructed by a food scientist rather than a botanist, we have selected (and improved through centuries of breeding) a subset of the plant kingdom for food use, to be valued and cultivated accordingly. Then, for each of the members of this subset, we have selected certain parts for consumption, following certain operations (cooking, boiling, chilling) to preserve, enhance, or make less hazardous the experience of consumption.

A similar categorization has of course been applied to the animal world, where animals are used for food directly (meat) or indirectly (milk, eggs, etc.), or else are domesticated or remain wild and uncooperative.

Then, just as we have evaluated which parts of the plant are most useful for our foodstuffs, we have learned the value and different properties of all the different parts of those animals we regard as "food." Our food-focused anatomical knowledge of the cow, for example, tells us what pieces will yield what kinds of meat products, probably being second only to our more medically focused anatomical knowledge of the human body. We have detailed understanding of the processes that take place after death for both medical- and meat-related reasons, in the latter case with the goal of being able to control the ultimate tenderness of a cooked piece of meat by careful actions during the critical few hours after death and the onset of rigor mortis.

A quick Google search will turn up many different versions of what are essentially anatomical charts for a cow, where we can see that few parts of the animal are not used for meat in one way or another, while the tougher parts like those around the neck or legs might give us meat best suited to slow-cooking methods like stewing, and the middle of the animal, loin and back yield the most tender and valuable steaks. The pig and sheep can be similarly viewed in such cold functional terms, where function in the body is related directly to

---

[15] They just find a tree that seems like a fun guy to be around and don't need much room among his roots to make their homes.

texture, palatability, and the best approach to transformation into cooked meat, as are birds such as the chicken, where the difference between meat found in active parts of the body (like the wing and leg) are clearly different to that in the less-active breast. In such cases, we can typically see the contrast between regions much more easily than for larger animals, for which the consumer normally only sees the part they are interested in cooking.

Again, these kinds of understanding of how to bend the living world around us to our nutritional and survival needs have been a critical part of humankind's success and ascent to dominance of the world's life-forms. As I will explore in the chapters to follow, every aspect of food, from the accumulation of tasty sugars in a peach to the difference between a sirloin steak and a piece of soft salmon, can be explained in terms of molecules and their properties. What happens when we take those raw materials so plentifully provided by the bounteous earth and prepare them for our consumption is likewise the result of the application of chemistry, with an occasional sprinkling of physics.

A key transformation that thus occurs between the state in which many raw materials of food are found in nature and the state ("food") in which it enters our mouths is (normally) that from a state of life to a state of nonlife, but where the latter state enables, enriches, and permits our own life. The processing of food in this way could be said to be a process of transfer of life from the natural world to us, and so life itself is the secret ingredient in all food products.

Even looking at food this way, though, we can see that food is teeming with life in many different ways throughout its journey. It can be assumed that, once harvested/caught/plucked/slaughtered, most "pre-food" will begin to deteriorate through a range of mechanisms, and much of our focus in this book will be on the means by which mankind has learned, at first empirically (by a harsh variant of trial and error called trial and death) but later more scientifically, to delay such undesirable processes.

These mechanisms by which the quality (whether in terms of taste, nutrition, safety, or attractiveness) of food is lost can be divided very broadly into two classes: those that are due to the tendency of living materials to continue the reactions that they engaged in while still alive, and those that are due to the presence of other forms of life, who hitched a ride either while the food material was in its original form or later on.

In the first class are found changes as mentioned earlier such as the ripening of fruits and similar changes in vegetables, which can be directly linked to the continuance of reactions that would have happened pre-harvest.

In the second class, we now know that, while we can see food with our eyes, other forms of life which we cannot see are to be found on almost every imaginable food product. These microorganisms, such as bacteria, molds, and yeasts, grow, move, and multiply across every surface of our food, and often through its interior, at numbers that are so huge that they are beyond our intuitive understanding. To put this in perspective, an average forkful of raw food might contain more bacterial cells than the populations of most European countries. Similarly, it has often been said that the human body contains around 10 times more bacterial cells than human cells, with most of those being found living in a harmless (and probably, under normal circumstances, beneficial) cohabitation arrangement in our digestive systems.[16]

To combat these changes, and make our food retain its most desirable properties for longer, we have learned to add things to food, or treat food in such a way as to encourage the development of useful preservative molecules (fermentation) or to subject food to external processes such as heating, cooling, or drying. All these, and their impacts on the chemistry and biology of food, will be discussed in the chapters to come.

---

[16] Assuming they do outnumber us 10 to 1, though, and we better hope they never work out how to vote; we do know that, if they ever decided to leave us (Bugsit?), we would be infinitely the poorer in terms of our health and basic survival.

{ 3 }

Consistency and Change
PROTEINS IN FOOD

Why are proteins important in food?

Proteins are, in my view, the most impressive molecules in food.

They influence the texture, crunch, chew, flow, color, flavor, and nutritional quality of food. Not only that, but they can radically change their properties and how they behave depending on the environment and, critically for food, in response to processes like heating. Even when broken down into smaller components they are important, for example giving cheese many of its critical flavor notes. Indeed, I would argue that perhaps the most fundamental phenomenon we encounter in cooking or processing food is the denaturation of proteins, as will be explained shortly.

Beyond food, the value of proteins and their properties is widespread across biology. Many of the most significant molecules in our body and that of any living organism (including plants and animals) are proteins. These include those that make hair and skin what they are, as well as the hemoglobin that transports oxygen around the body in our blood.

What exactly are proteins anyway?

Proteins are built from amino acids, a family of 20 closely related small molecules, which all have in chemical terms the same two ends (chemically speaking, an amino end and an acidic end, hence the name) but differ in the middle. This bit in the middle varies from amino acid to amino acid, from simple (a hydrogen atom in the case of glycine, the simplest amino acid) to much more complex structures. Amino acids can link up very neatly, as the amino end of one can form a bond (called a peptide bond) with the acid end of another, and so forth, so that chains of amino acids are formed that, when big enough (more than a few dozen amino acids), we call proteins.

Our bodies produce thousands of proteins for different functions, and the instructions for which amino acids combine to make which proteins are essentially what the genetic code encrypted in our DNA specifies. We hear a lot about our genes encoding the secrets of life, but what that code spells is basically P-R-O-T-E-I-N. Yes, these are very important molecules!

The simplest level of structure of a protein (called the primary structure) is that sequence of amino acids, which can be written out as a (long) line. However, this doesn't mean that protein molecules are all long, skinny chains of amino acids, as far more complex structures arise from the nature of the side chains attached to the amino acids along the way.

For example, some of these side chains attract each other (chemically speaking) while others repel, and local interactions of this sort can introduce kinks or twists into the chain accordingly. Also, some amino acids or stretches of these like water, while some utterly detest the stuff. For this reason, proteins will arrange themselves to be most stable in the aqueous environment in which they are most frequently found so as to have the former on the outside of a three-dimensional structure while the latter is buried in the interior of that structure. Finally, and very important for the properties of proteins in food, one sulfur-rich amino acid (called cysteine) doesn't like to exist alone because of its highly reactive side chain (called a thiol group) and, when two of these amino acids are in a protein, they tend to form a very strong bond between themselves, called a disulfide bond. If these cysteines are relatively well spaced away from each other along the amino acid chain, this can result in the protein's effectively tying itself up in knots. In proteins, disulfide bonds (so-called because each cysteine contains a sulfur atom) form a key part of the protein structure, and cysteine residues are usually found in multiples of two in proteins, giving half that number of disulfide bonds.

So, to sum up, the DNA code in cells describes the sequence of proteins, and the cell machinery follows these instructions and produces the proteins, which then assemble into their unique three-dimensional shape based on their environment and the unique sequence of amino acids, and their preferences for associating with each other, how they react with each other, whether they attract or repel their neighbors, and a host of other factors. Simple, right? Perhaps not surprisingly, predicting protein structures by taking into account all of these factors was long reckoned to be one of the most difficult problems in science.

It is known, however, that the function of proteins depends critically on their structure. Hair is made of protein, but the nature of someone's hair, whether straight, curly, or untameably frizzy, and how it responds to water and shampoo, all depend on their structure and ultimately the properties of the amino acids that make them up.

Likewise, a very special category of proteins, especially in food, is enzymes, which are proteins that have the ability to act on other molecules and exert specific changes; for example, when we eat food, the molecules are broken down in our body to release the required nutrients by a series of enzymes. There are even cannibalistic enzymes that break down other proteins, called proteinases or proteases.[1] We will talk more about enzymes in food in several places in this book, because they are very very significant.[2]

So, that was a very brief description of where proteins come from, how they fold, and how structure defines their function. Now, how does that relate to food?

As we have seen, most food products come from plant or animal sources and, just as those sources contain a range of proteins, so do the food products derived from them, and we rely on their amino acids for nutrition (especially critical are the so-called essential amino acids, which we must consume because our bodies cannot produce them naturally). I have already touched on how important proteins are in terms of the life of everything on earth, how many fundamental processes in our bodies are dependent on proteins, and how one of the key reactions of life is the building of proteins from instructions encoded in DNA. The raw materials for making proteins are amino acids, but our bodies cannot produce all the amino acids we need, and so 9 of the 20 most common amino acids (such as tryptophan and lysine) must be got from the diet, giving proteins a critical nutritional function. Food proteins are often classified in terms of the efficiency with which they provide a complete and balanced amino acid intake, and eggs, meat, and soybeans are known, for example, to provide all essential amino acids. Other important nutritional properties of proteins are their ability to carry and transport into our bodies certain critical nutrients (such as the calcium–milk protein we will discuss shortly) and the way in which they are digested in our bodies to best release their amino acids.

However, besides just providing essential amino acids and other nutritional benefits, proteins do much more besides and are responsible for many of the critical properties of food products and. In particular, they are responsible for the effects seen when we take foods and their molecules outside their comfort zone and expose them to conditions that they would never see in their natural environment, whether that be by mixing with other molecules, adding acid, or applying high temperatures when we heat or cook the food.

To give an example of the complexity and contribution of proteins in food, I am going to focus on the milk protein system in this chapter, before moving on to meat and other systems in the next.

Two tribes: The milk protein system

Milk contains around 35 grams of protein per liter, and the proteins in milk fall into two groups with very different properties; around three-quarters of the protein by weight is called casein, and one-quarter is called whey protein.

---

[1] This process is called "proteolysis," from two conjoined roots, "proteo" and "lysis," and this is for some reason one of my favorite scientific words, along with its adjective form, "proteolytic."

[2] Conflict of interest declaration: I have studied food proteins and proteolytic enzymes for 25 years, so I may be just a little biased

Separating proteins in science often requires very complex laboratory equipment and challenging methods; separating these milk protein families can be done in any kitchen. This is because these families exist in a relationship a bit like that of housemates who don't get on with each other, so that the slightest stress can encourage them to get as far away as possible from each other. In this case, the stress is as simple as adding vinegar.

Take a glass of milk (skim milk is easiest) and add vinegar drop by drop, and at some point the milk will curdle, with a lump of white solids settling to the bottom of the glass, leaving behind a translucent greenish liquid. The casein is the settled stuff, and the clear liquid contains the whey proteins, presumably delighted to be rid of their unstable housemates, even if the house smells a little as a result.[3]

Why does this happen?

The protein family known as casein in fact contains four major related proteins, called by the Greek names beta, kappa, α_{s1}, and α_{s2}.[4] These proteins share some basic features, being relatively simple unfolded strings of amino acids, rich in certain amino acids such as serine (of which more in a minute) and lysine. They are also mostly not very fond of water, something of a disadvantage in a system like milk, which contains almost 90% of the stuff. To exist in this hostile environment, they huddle together (molecularly speaking) into structures that minimize their contact with the dreaded water, while one of them (kappa) is delegated to playing a heroic role in keeping them safe and stable.

How does it do that? Kappa is somewhat of a two-faced molecule, having a water-unfriendly (or hydrophobic, as in phobic of hydro, from the Greek for water) part, and a hydrophilic (as in the opposite of hydrophobic, and thus water-loving) part, called the caseinomacropeptide. This water-loving property is conferred by the fact that this bit of the molecule is covered in a special type of sugar molecule, and sugars, as we will discuss in Chapter 6, are usually very water friendly. So, kappa finds its favorite place, like a dog finding the comfiest spot on the couch, right at the interface between its brethren in the casein family and their dreaded watery surroundings.

The most stable way for this to happen, which is nature's way of minimizing the surface area of contact between casein and water, is for the caseins to be found in spherical aggregates, called micelles, with the kappa keeping peace at the surface. Each micelle contains thousands of molecules of the different caseins and can easily be seen under powerful electron microscopes (we will see them in some

---

[3] For many scientists, isolating proteins is a very difficult task, and obtaining reasonable quantities for their studies can be a very laborious consideration. Such scientists must be a little jealous that dairy scientists can isolate their main topic of study relatively easily in a kitchen in large quantities with stuff they bought in their local supermarket.

[4] It was originally thought there were only three, until more advanced methods revealed that the alpha fraction was in fact two proteins, subsequently renamed s1 and 2 (the 's' referring to their calcium-sensititity).

images discussed later in this book). Zoomed in using the extremely powerful magnification, micelles look like tiny cauliflowers, with wrinkled surfaces from which (although we haven't been able to peer closely enough to see them) the kappa casein molecules are believed to protrude, like hairs or the bristles on a brush, both terms that scientists actually use to describe this protective layer.

Sounds complicated? We are only getting started here.

Holding the micelles together requires other forces, and one of the critical ones involves one of the most famous parts of milk, which is calcium.

Growing up in Ireland, a very well-known ad on TV for dairy products bore a fiercely annoying, both in grammatical and musical terms, tune that went "them [*sic*] bones, them bones need calcium." The message persists; milk contains calcium, and a lot of it, just what them bones need. However, if we had a certain volume of water and tried our best to dissolve calcium (specifically the form of calcium found in milk, called calcium phosphate) into it, we couldn't manage to fit in as much by volume as we find in milk, and beyond a certain point it would become insoluble and fall out of the liquid as a solid deposit.

How milk manages to contain so much of this bone-boosting mineral is related to the casein micelles, as the extra calcium is dispersed and bound inside these protein structures, which thus act as tiny calcium-delivery vehicles, dispersing the calcium in infinitesimal crystals throughout the milk, too finely divided to fall out of solution.

In return for this molecular piggybacking, the calcium confers structural stability on the casein micelles, acting like a sort of glue to hold this loose assemblage of thousands of protein molecules together in their spherical homes. The calcium phosphate molecules are bound to amino acids in the caseins called serine, but can link to more than one, and so can draw neighboring casein molecules together and hold them together within the micelles.

The exact structure of the casein micelle has been a topic of perhaps unexpectedly heated debate within the international community of milk protein scientists, and remains today somewhat unclear, but it is generally accepted that the casein molecules within the micelle are tangled together like a ball of wool strands scrunched up, with the calcium phosphate crystals forming bonds holding them together, and the kappa casein then being found on the outside.

It is in this form that micelles are found in milk fresh from the cow, and remain largely unchanged throughout the processing of milk into what we purchase from the market.

When I have said that the micelles are tiny, we should put that in perspective. Each micelle is around 200 nanometers in diameter, or 0.0002 millimeters across. Each drop of milk contains millions of these micelles, and each micelle contains thousands of casein molecules wrapped up in a loose ball, containing quite a lot of water. I have often wondered what a casein micelle would feel like, if either a person was shrunk to sufficient size to hold one in their hand, or else the micelle expanded to the size of a tennis ball (which would make some very interesting cheese); a general view is that it would feel like a sponge, but with an itchy hairy surface.

With such genuinely microscopic particles, we can't of course see the micelles directly, but the white color of milk shows them to us indirectly. Milk doesn't contain anything that colors it white, and its appearance is actually due to its playing some tricks on the light that passes through it. When there are particles of a certain size in a solution, they interfere with (scatter) the light passing through (such particles are referred to as being colloidal). In raw milk from a cow or whole milk in a commercial container, the particles that do most of this scattering are fat globules (of which more in Chapter 6). Skim milk, from which almost all the fat has been removed, also looks white, though, and this is due to the casein micelles having a similar effect. If we add agents that cause the micelles to fall apart (the secret to this is often grabbing the calcium glue and sucking it out of the micelles), the whiteness disappears, and the milk looks almost translucent, or is in fact a light green color (because of the presence of a colorful vitamin called riboflavin). This is also what the milk looks like if we remove the casein, as in our earlier stinky experiment with the vinegar.

The casein system is thus very delicate and can easily be destabilized and changed, sometimes in a very small way that mightn't be noticed and sometimes in a very significant way that really can't be missed (as in the case of vinegar).

One more subtle change results from the simple act of keeping the milk cold, which weakens one of the other forces that keeps the caseins in micelles, called hydrophobic interactions. When these are reduced, one of the caseins, beta, feels a lot less fraternal love for its family members than before, and some of it, although usually found within the core of the micelle at warmer temperatures, tends to wander off into the surrounding watery milk phase, past the protective kappa casein fence, like a sulky teenager sneaking out of the house. This doesn't have much noticeable effect, however.

Much more noticeable things happen when we introduce acids like vinegar into the system.

The magical influence of acid, and the accident of yogurt

Part of what keeps the micelles apart in milk is their electrical charge and the acidity of milk, which is measured in terms of a property called pH that reflects the extent to which any material is acidic or alkaline. This scale runs from 1 (incredibly strong acid) to 14 (incredibly strong alkali), and milk usually has a pH just on the acidic side of the neutral middle of this range, around 6.7. Amino acids, depending on the chemistry of the side chains mentioned in section "What exactly are proteins anyway?" can bear different electrical charges at this pH, some being electrically negative and some positive. Overall, though, the balance at the pH of milk is toward negative charge, and so the micelles are negatively charged overall. A basic principle of physics (electrostatic repulsion and attraction) explains that particles or molecules (and large assemblies of molecules, like the casein micelles) that bear the same electrical charge will physically repel each other, while those that bear opposite charges will attract.[5]

In the case of milk and micelles, this means that, while large negatively charged micelles float around in milk, when they come too close to their neighbor the electrical charge effect will push them away again, so meaning that the micelles remain suspended and don't stick together.

Unless . . . here is where the vinegar comes in.

The balance between the negative and positive charges depends on the pH, as just stated, and as the pH changes the balance changes as well. When we add the vinegar, the pH of the surrounding medium becomes lower (more acidic) and the negative charge on the micelles becomes less and less and less until, at a critical value of 4.6, it becomes zero (this is called the isoelectric point of the protein). Thus, one of the main forces keeping micelles apart has just been canceled, and we now have proteins in an unstable state that, overall, don't much like the watery environment they are in. To stabilize themselves, they try and maximize contact with each other and minimize contact with the water.

In other words, they cluster, clump, and form structures such as chains and three-dimensional networks, all with the objective of maximizing micelle–micelle contact (so long forbidden by the now-banished tyranny of electrical repulsion) and minimizing their contact with water. This is one of the most amazing properties of any molecule I know in food and allows casein micelles to form the basis of many food structures, like yogurt and cheese, that we encounter every day.

On the one hand, if the change in environmental pH happens very quickly, like when we add the vinegar, the destabilization and interactions happen in a very chaotic way, hence the precipitation into the lumps at the bottom of the container.

On the other hand, if the change is very slow, and the micelles gradually form their new links under gentle undisturbed conditions, the milk solidifies into a gel. This was probably first observed some millennia ago, when it happened by accident because of milk being left exposed to air in warm conditions, after which a curious observer of this phenomenon must have felt sufficiently intrigued to taste the result and found it to be refreshingly acidic. Thus was yogurt discovered, a long time before the underpinning science was understood.

This science involves a number of types of bacteria, found commonly in the environment, which like to ferment the sugar in milk, lactose. These bacteria are luckily relatively benign and don't cause any health problems, but their consumption of lactose produces a by-product called lactic acid, which reduces the pH of milk as it is produced. Typically, it takes a few hours of favorable fermentation with such bacteria for the pH to reach the magic value of 4.6, and this rate is just perfect for the micelles to design their most complex and interconnected network, and for the milk to solidify into the yogurt gel. In a modern yogurt plant, there is a bit more

---

[5] This principle is exploited in some familiar tricks like causing balloons to stick to walls if first rubbed on your hair, which is all about transferring charges to get attraction to occur where you mightn't expect it.

complexity to the process than this, but the heart of it remains: exploiting these bacteria to ferment the lactose to acid and gradually transform the milk from a liquid into a soft gel-like solid.

This isn't the only way in which milk can be converted from a liquid to a solid though, and the stability of the casein micelles can be destroyed in a very different manner to yield a different type of dairy product, a phenomenon that has also been exploited for a very long time.

From milk to cheese

Somewhere,[6] a very long time ago, dairy legend[7] holds that a nomad had the brain-wave of transporting milk across distances, perhaps on horses or other animals, by storing it in what appeared to be a suitable, if somewhat distasteful by modern sensibilities, bag-like container: a calf stomach. Holding milk at warm temperatures, along with the inevitable agitation caused by the motion of the bag, resulted in a puzzling transformation of the liquid milk into a mixture of soft solid curds and a surrounding watery liquid. Again, curiosity or hunger must have driven someone to taste this odd mixture, and it was discovered that the curds, particularly if pressed together to squeeze out some more of the liquid, could form an interesting new product, which we now call cheese.

The key factor that resulted in this presumptive accidental discovery was the fact that calf stomachs contain an enzyme that has a unique action on milk. This remarkable enzyme is called chymosin, and a chymosin-rich extract from calf stomachs has been known by cheesemakers for decades as rennet.[8]

The action of chymosin can destabilize the casein micelles in such a way that, if added to milk at warm temperatures, without agitation, it will on its own (no bacteria needed, no acid produced) transform liquid milk into a solid gel that, when cut and stirred and then pressed to exude a watery liquid from which most of the fat and protein in the milk have been depleted, gives the base curd that can become cheese, while we call the separated liquid whey.

The mechanism by which chymosin can do this is related to the stabilizing part of the kappa casein molecule mentioned before, the sugar-rich caseinomacropeptide. Chymosin is a proteolytic enzyme, and it snips a range of peptide bonds in the caseins, chewing the intact proteins down into smaller polypeptides and peptides. Indeed, many enzymes do this (as we will see in the context of cheese ripening in Chapter 4) but what makes chymosin significant is that, out of all the bonds it can act on, it has a huge preference for one that essentially holds the caseinomacropeptide

---

[6] Perhaps in a region that could be called Curdistan

[7] Dairy legends tend to start, "once upon a time, in a place far far a-whey . . . ."

[8] Rennet also contains another enzyme, called pepsin, also found in human stomachs, and so the two terms aren't interchangeable.

(and its stabilizing sugars) onto the rest of the kappa casein molecule, and hence in effect onto the micelle itself.

When chymosin is added to milk, it attacks this bond with a single-minded fixity of purpose, and in effect gives the casein micelles a haircut, shearing off its single greatest stabilizing influence. As a result, the micelles are suddenly stripped of that which made them even slightly comfortable (in an inhospitable aqueous environment), and again their natural instinct is to seek solace in each other. So, they huddle together in their chains and networks,[9] and the milk turns (in a typical cheesemaking scenario within around 30–40 minutes) into a semisolid gel with the consistency of weak jelly, which can easily be cut with a knife or blades into soft squishy curds. These are then stirred in their surrounding whey, and usually cooked by heating them up by 8–20 °C, depending on variety, to expel more whey, toughen them up, and then recover them and press them into a block or cylinder, to which a little salt is added, to give the precursor of what we will come to call cheese.

Chymosin actually isn't the only enzyme that is able to coagulate milk in this way, and lots of others from plant, microbial, and other sources have been found to do the same trick. The reason chymosin is the king though is its almost obsessive attention to the one critical bond that will destabilize the micelles. Other enzymes will also attack different peptide bonds in the caseins while the milk is clotting, and each of these attacks weakens the gel, reducing the amount of intact structural material, and chews off little bits of the casein that get lost in the whey rather than remaining in the curd.

Perhaps not surprisingly, in recent decades there has been a move away from coagulating milk with an extract from calf stomachs, which is one reason why many alternative coagulants have been investigated, but generally found wanting. The most common solution reached is to use chymosin that is produced by bacteria or yeast in very controlled conditions (called fermentation-produced chymosin). How this is achieved is a marvel of modern biotechnology, as the sequence of DNA that codes for the instructions of how to make the enzyme has been identified in the cow genome, and then an exact copy of this inserted into the genome of the host microorganism in such a way that it is produced alongside all the normal proteins that cell routinely manufactures, and secreted from the cell into its surroundings. Then it is relatively simple to recover it from the fermentation vessel and use this for cheesemaking.

The outer and inner structures of the complex dairy products that can be made from exploiting the coagulation properties of casein are shown in Figures 3.1 and 3.2. In the first picture, we can almost smell the rich, lactic pungency of the aged Parmesan, coagulated from milk and then allowed to ripen for probably more than 6 months, in which time the protein coagulated by the action of rennet has broken

---

[9] Calcium is needed for this to happen, in another example of the complex and very important relationship between calcium and the milk proteins. The temperature also can't be too low, as renneting milk in the cold will result in milk that stays a liquid, but gels quite quickly when warmed up.

down (as has the fat and lactose) to transform the bland starting curd into the complex texture and flavor that we taste and admire.

Peer once again below the surface though, and the complexity of flavor notes is found to be matched by complexity of structure. Any fermented dairy product, such as the blue cheese and yogurt seen in Figure 3.2, shows a characteristic network of coagulated casein micelles, linked into chains, nodes, and bridges, which give the solid structure we see quite clearly in the yogurt sample shown, alongside the causative agent of this coagulation, a single cell of the lactic acid bacteria (*Streptococcus*, by its shape) seeming to hover like a spaceship over the fruits of its fermentative gusto. In the case of the blue cheese sample, aged for many months, the details of the casein network are far less clear, perhaps because of its progressive breakdown, and in this case the live guest star is the thread-like hyphae of the blue mold (likely *Penicillium*, as the name suggests related to the mold that gave us penicillin), growing wild over the protein network and leading to the appearance of a forest floor dramatically overgrown with some creeping, brambly weed. The potent complement of enzymes from such molds are what give cheese like Stilton and Roquefort their strong peppery taste, as we will discuss more in the next chapter.

The remarkable behavior of whey proteins

As described in the previous section, milk contains two families of proteins, and we have focused our attention fully so far on the one that accounts for around three-quarters of the protein, the caseins. The remainder are the whey proteins which, as their name suggests, are found in whey produced after we coagulate or precipitate the casein by using acid or chymosin.

The whey protein family is more varied than the casein family, but the two biggest players are called beta-lactoglobulin and alpha-lactalbumin. The next most common ones are serum albumin and immunoglobulins, and then there is a regular zoo of minor whey proteins.

Of the whey proteins, probably the most interesting is beta-lactoglobulin, for a variety of reasons. This is a protein with a much more complex structure than the caseins, and it has what is called a globular structure, being found in raw milk folded up in a three-dimensional shape, which is held together by lots of disulfide bonds between cysteine amino acids. It also, however, houses a secret weapon, for it is one of the unusual proteins that has an uneven number of these amino acids, and so has one unpaired residue that isn't in a stable disulfide bond, and is deeply and angrily jealous of its siblings that are. The molecule senses this danger and folds itself carefully to hide away this angry radical in a cage within the molecule's core, far from where it can cause trouble.

This is how it is found in raw milk, in which the molecules of beta-lactoglobulin are normally fairly inert, sometimes forming small complexes of two of these molecules sticking together, but generally ignoring the caseins and other proteins present.

This is a deceptively peaceful vista, however, and all it takes is a little nudge to turn the Bruce Banner version of beta-lactoglobulin into a raging Hulk version; this nudge is one frequently encountered in processing milk, which is heat.

When protein molecules with a complex three-dimensional structure are exposed to increasing temperatures, the bonds that keep them in their usual (so-called "native") structure loosen or are broken, and the protein unfolds into a new, more open state. When the temperature decreases again, the protein might fold back into its original state, with no damage done. Sometimes, however, the protein completely changes its structure, adopts new structures, or interacts with other molecules. In such cases, when the heat comes off, what is left behind is very different to the protein we had at the beginning, and we refer to this as a "denatured" protein.[10]

Beta-lactoglobulin denatures with a vengeance.

When the molecule unfolds, at temperatures typically above around 75 °C (significantly, denaturation isn't much of a factor at the temperatures we use to pasteurize milk), the free cysteine residue and its highly reactive thiol group are released into the wild, and all that poor amino acid wants to do is to form a bond with another cysteine, like it has had to watch its brethren do until this point.

So, effectively, when heated, beta-lactoglobulin seeks with great ardor to form a new bond with any molecule that will have it, like a lovesick teenager.[11] The free cysteine residue will attack other disulfide bonds and try and join in, like someone tapping a dancing couple on the shoulder and stepping in, to start their own dance, and perhaps forming a new bond, as it were.

This new bond could be with other molecules of beta-lactoglobulin, leading to polymers or structures composed of multiple individual molecules of the same protein, which can end up being very large. It could also be molecules of other whey proteins.

It could also be the casein micelle, however, as kappa casein, last seen stabilizing the surface of the micelle, holds one of the prized disulfide bonds. So, in strongly heated milk, the casein micelle is found to have grown a whole new hairy layer (hair extensions, as it were), with denatured beta-lactoglobulin coating the surface. This is one reason why milk for cheesemaking is never heated too severely (past the minimum pasteurization conditions needed to make it safe), as such a coating makes it very hard for chymosin to do its job and for the gel to assemble correctly, with this messy, badly behaved protein stuck in the way (when it should be in the whey instead). In contrast, such interactions are very important when we are making yogurt, and to be welcomed, because the acid gel that results when whey proteins

---

[10] Isn't "denatured" is great word? It literally means taken away from the state that nature intended. Nature clearly didn't think about what would happen if cows had a body temperature above 75 °C.

[11] It is hard when discussing proteins to avoid a certain molecular anthropomorphism, and attribute human characteristics to their frequently eccentric behavior.

have denatured and interacted with the caseins is stronger and less likely to contract during storage and expel whey (giving an undesirable watery layer on the surface of the yogurt in the pot).

These properties of beta-lactoglobulin make it a very valuable food ingredient, and are a reason why whey ingredients appear in a wide range of food products we see every day, and the story of how this came to be is the result of one of the most remarkable transformations in the food sector.

As one final observation of beta-lactoglobulin's interesting properties, it is absent from human milk, unusual among mammalian species. In addition, this protein can cause allergic reactions to milk, and so removing it or otherwise making it less allergic (by using enzymes to chop it into harmless tiny fragments, in an equivalent process to destroying an incriminating document by passing it through a shredder) may be a key consideration in the production of infant formula for babies.

Whey processing: Never waste a good crisis!

For every 1000 liters of milk that come into a factory to be transformed into cheese, 900 liters are left behind in the form of whey.

The casein and most of the fat have been concentrated into the cheese curd, and what is left behind is a watery solution with around half the level of solids found in milk, being mainly composed of the milk sugar lactose, with less than 1% protein (whey protein) and some soluble milk salts.

For centuries, this was regarded as a useless by-product, and essentially a problem of disposal onto land, as feed for pigs, or dumped into nearby waterways. A dramatic photograph I once saw showed the dismantling of a large pipeline that carried millions of gallons of whey from cheese factories directly into the sea.

In the 1970s, however, concerns from a wide range of perspectives about the wisdom of such disposal led to interest in whether whey could be more usefully handled, and whether food ingredients could be produced from whey. In terms of recovering the proteins specifically, the main challenge is to get rid of all the water, but if whey is simply dried the resulting powder is mainly lactose, with less than a fifth being protein, and adding that to any food product means adding a lot of lactose for every bit of interesting protein.

So, a range of technologies was developed, mainly in Ireland and France, which involved filtering the whey under pressure through membranes (essentially plastic filters, which will be discussed more in Chapter 12) that retain the large protein molecules, while allowing the much smaller sugar and salts molecules to pass through. The key process used here, called ultrafiltration, essentially purges the protein of the less-desirable fellow-traveler constituents of whey, and, after drying what does not pass through the membrane (called the retentate), products with progressively higher levels of protein can be obtained, leading to a family of products called whey protein concentrates or, when the process is modified to obtain very high protein purity, whey protein isolates. Other modifications to the process, which manipulate conditions so as to exploit differences in properties of the individual proteins under varying environmental conditions, can even give products enriched in individual whey proteins, and even purified beta-lactoglobulin. The remaining by-products of such processes are not wasted, with much of the lactose, for example, ending up being used as filler in pharmaceutical tablets or being converted or fermented into other useful products.

In this way, milk coming into a cheese plant can ultimately end up producing a whole range of different products by stripping the raw material for parts, as it were. It is frequently joked that, whereas whey was once the low-value by-product of making cheese, cheese is now the low-value by-product of making whey.

But why are whey proteins so valuable and interesting that they merited such efforts to make them available? One reason is the structural changes caused by their denaturation that I previously mentioned.

As an example of this, purified whey proteins can be readily dissolved in water at quite high levels, and a solution of 10% whey protein, at the right conditions of salts and pH in the solution, will essentially look like water. If this solution is heated carefully, however, up to around 90 °C for even a few minutes, it will magically transform into a solid material with a thick gel-like consistency. This is the result of all the molecules of beta-lactoglobulin unfolding under the heat, exposing their reactive groups, and a process of complexing of individual modules into pairs, threes, and so on taking place, so that eventually there are enough links and large-enough aggregates of linked molecules that the system solidifies.

When considering this behavior of beta-lactoglobulin, as with the behavior of casein in forming structures as discussed earlier in this chapter, it is hard to resist the analogy of molecular Lego blocks. In milk, the micelles and whey proteins are raw blocks, disordered, and with no real interaction, but apply the right conditions (heat, acid, an enthusiastic child, as the case may be) and these tiny blocks suddenly attain a new purpose and create a wide range of outcomes.

Having a glass of liquid that you can turn into a solid by heating is of course a neat party trick[12] but, at lower protein levels, or under different circumstances, such interactions and structure-building ability can thicken, strengthen, or otherwise contribute to the texture of a wide range of food products, from sauces to yogurt and meat, and this is why whey proteins have ended up in a wide range of food applications far from their original home.

Consistency and change in food: The roles of proteins

Overall, these examples of the power and properties of milk proteins have hopefully illustrated how important they are in determining many properties of products we know and love.

---

[12] The reception to which might depend a little on the type of party at which it is performed.

As the title of this chapter said, they are responsible for the consistency of food products such as cheese and yogurt, and effectively dominate the structure of these, but also are dynamic entities subject to change when exposed to different conditions and treatments, and understanding how these affect the structures of the proteins and the products in which they are found is critical.

One thing that didn't appear much in this chapter of course is flavor, and protein could perhaps be seen as the big dumb scaffolding that gives food its texture, while much more interesting molecules are responsible for flavor, aroma, and all the things we love about our food products. I prefer to think that proteins create the canvas on which flavor is created and give food its strength, while even, particularly after they are broken down, undoubtedly contributing to some flavor characteristics, as we will explore in later chapters.

In the next chapter, we next consider other proteins besides the milk system and see how a lot more than dairy products rely on proteins for their properties, their consistency, and many of the changes we see and take for granted.

{ 4 }

Build 'Em Up and Break 'Em Down

When proteins break down

Proteins are not just interesting and significant in food in their intact or even aggregated or complexed states, but often lend their greatest value to food by their disappearance.

For example, in cheese, as we discussed in the last chapter, proteins are critical for the coagulation of milk and conversion into curd, and cheesemakers choose the enzymes they use to cause that coagulation specifically for their lack of other impact on the milk protein casein. However, once the cheese is made, the intactness of the casein abruptly becomes a liability, in a sort of cosmic ingratitude, and indeed the cheese will not be considered fit to eat until it is at least partially gone.

The reason for this is immediately apparent if the freshly made cheese (of just about any variety) is tasted. Does it taste like cheese? Only if you like your cheese bland and flavored almost entirely of salt. Does it have the texture of cheese? Only if you think cheese should take quite a while to chew while savoring its boring salt-iness. This is the taste and flavor of (salted) intact casein curd.

So, no one eats cheese in this state, and almost every variety of cheese, from Accasciato[1] to Zamorano,[2] is held for at least some time after manufacture to undergo what is called ripening,[3] during which it develops the flavor and texture we will expect it to have.

In the case of the Parmesan shown in Figure 3.1, the crumbliness we associate so closely with this cheese is achieved by a combination of breakdown of the protein network over very long ripening times (often 12 months to 2 years), and a parallel drying out to low-moisture contents (which also concentrates the wide range of

---

[1] A soft creamy Italian cheese.

[2] A Spanish sheep's milk cheese.

[3] An interesting term, somewhat placing cheese as an analogue of a fruit, which needs to change to reach an acceptable state of maturity; we don't refer to cheese as "ripe" when the process is complete, though, switching metaphors abruptly into instead terms relating to maturity.

compounds produced during such long ripening to give a very strong and piquant flavor).

Interestingly, almost all freshly made cheeses enter the ripening stage with similar flavor, but they leave with a ridiculously wide variety of characteristics. What gives rise to these differences? Some of the differences between varieties arise from the type of milk, be it from cow, goat, sheep, or buffalo, which brings in subtly different flavor notes, or more often the precursors of these, mainly derived from the different feeds of the different animals. More arise from the environment of the ripening area (Is the cheese packaged? What temperature is it at—typically in the range 8–20 °C? Is the air humid? Is it in a cave or a steel-walled room?)

Finally, the most profound differences arise from the biological agents we unleash on the fresh milk constituents trapped in the curd, which we entrust to transform it into the desired end product. Specifically, these include the coagulant used and the bacteria or other microorganisms (like yeasts and molds) we add to the milk, add to the cheese, or encourage to grow, sometimes more in hope than with precise control.

What all have in common though is the breakdown[4] of milk constituents such as the proteins. The milk itself, the coagulant, and the microorganisms present collectively possess a powerful arsenal of weapons of cheese-mass destruction, called enzymes, which we first met in the form of rennet in the previous chapter.

These enzymes act like scissors, and each has a preference for different parts of the casein molecules. Each of these molecules contains around 200 amino acids linked together and some enzymes like one or more of the specific bonds between these links in the protein chain. Some will act on large chunks of protein, like the intact casein, or very large fragments thereof, and work like axes or chainsaws to chop it up into slightly smaller chunks. Then, other enzymes might take over and act preferentially on these chunks like shears to chop them down further, while yet others will prefer progressively smaller chunks, and act like fine pruning scissors to finely dice them up, eventually into short assemblies of two or three amino acids (dipeptides or tripeptides, respectively) or even free amino acids. These then get further broken down to a range of different products by yet more enzymes.

So, week by week, month by month, as the cheese is held in the ripening room, these enzymes work together, sometimes competing for their preferred raw material (substrate), in other cases waiting patiently for a previous step to be complete and essentially feed them. Week by week, month by month, the protein is slowly chewed up, and its structure breaks down, leading the cheese to soften, while the overall flavor profile becomes progressively more complex, as more flavor-contributing products appear. This broken down structure is seen in the images of blue cheese in Figure 3.2.

---

[4] The term "degradation" is often used for such changes, but I do not like this and will avoid it because of, for me, the connotation of a change to a less-favorable state.

If samples of Cheddar, Edam, Camembert, and a piquant Blue were analyzed at their point of optimal flavor (were it easy to define such a thing), they would, although they started with an identical set of caseins, have very different levels of residual intact caseins and quite different profiles of the products thereof, right down to amino acids and beyond; each profile could be traced back to the agents (particularly added "starter"[5] bacterial cultures) added and ripening conditions applied.

Describing the methods we can use to study the actual proteins in food products such as cheese and to follow their breakdown and interactions gets us into a significant level of scientific detail, but for those interested the key principles of the methods concerned are explored in the Appendix.

Proteins in meat and fish: Muscles and mussels

As described earlier in this book, proteins are my favorite part of food, because I have studied them for over 20 years, and they do a lot of cool stuff without fuss or fanfare. The gel of yogurt, the chew of a steak, the coagulation of a fried egg—these are all thanks to proteins.

We know that a fillet steak and a fillet of salmon have very different textures. To understand this, we need to think a little more about the difference between a cow and a fish, and what they spend large portions of their day doing. To either stand or float and swim requires muscle, and meat is muscle. However, the structure of that muscle will reflect the demands of gravity and the need to overcome them to stand upright, with or without the buoyant support of water. Thanks to the intrinsic support of water, fish need far less strong muscles, and so different kinds of strength-conferring proteins, which is directly reflected in what we detect when we bite into them. In addition, fish don't need a lot of connective tissue to support their muscles, so we don't need to cook for ages like we might if this were present, and so fish can cook much more readily than land-borne meat.

In the case of the structure of meat and fish, as stated, one of the key factors affecting muscle structure (which then directly affects meat texture) will be how much exercise an animal undertakes, and how much weight it needs to support itself while doing this. This determines the large-scale differences between meat and fish, and even between different types of meat-yielding animals, such as chicken, pork, and beef, depending on how active the animal in question was during its life.

Even within a single animal, the functions of different parts of the body, in terms of what exercise they undertake and what their function is, mean that they will give meat of different characteristics and texture. In a chicken, for example, legs and wings will be active and require active muscle to do that they need to do, while the breast is essentially a nonmoving part, serving as a fuel store that doesn't need

---

[5] Starter bacteria are those deliberately added to milk by the cheesemaker, and the term captures how they play a key role at the start of the process, by producing acid and kick-starting the ripening process.

the ability to move rapidly and respond quickly. Biology will design any animal for function, and each part will be constructed in line with their intended activities. This means that different amounts and types of proteins will be found in different parts of the bodies, as well as different levels of energy-storing or -releasing sugars and fats, and so the meat that is obtained from different parts of that animal will differ from body part to body part.

Whatever the source of the meat, the basic proteins present are similar, and in particular the effects of heat on these when we cook are the same. The main proteins in meat are called actin, myosin, and collagen. The difference between meat types can then often be expressed in terms of differences in the relative amounts of these proteins.

Sometimes, as we will see later in the case of gluten in bread, the key structural agent in a food product is not a single protein, but a complex of two or more, and this is also the case in meat, where actin and myosin work together to form one of the most efficient machines nature has yet devised. Indeed, as I type these words it is the concerted actions of actin and myosin that allow my fingers to move across the keyboard, keep my eyes moving to follow what I am writing, and even make my heart beat to make sure I am still alive to keep typing. These are clearly powerful and important molecules even before they confer on your steak the perfect texture!

Muscle filaments are composed of thick filaments of hundreds of molecules of the protein myosin, linked to thinner filaments containing actin molecules; the relative motion of these two types of filaments across each other is the basis of muscle movement. In chemical terms, myosin and actin can bond together at specific sites along the filaments, and such bonds can form and break spontaneously. When bound together, the filament network has one structure but, when they are unbound, has a different structure; these transient changes in the extent of interactions mean that the filament network can extend or contract, which, when many filaments act together, moves our muscles. Other proteins, including one called tropomyosin, are present to help coordinate this action, and can be found looped around the actin chains to make sure that extension and contraction occur smoothly and exactly when, and only when, they are supposed to, giving an exquisitely finely tuned biological machine.

There are also differences between parts of an animal (and thus pieces of meat) in terms of the specific types of myosin fibers and energy-providing mechanisms that are present. On the one hand, so-called "fast" fibers use glycogen as their fuel source,[6] while, on the other hand (or equivalent body part), "slow" fibers get their power from alternative pathways that use oxygen, and so need myoglobin present as a source of this oxygen. So, meat that is rich in slow fibers is darker and more red in color (thanks to the myoglobin) than that which is rich in fast fibers. Slow fibers are found in body parts that do not require rapid activity, but more ongoing strength

---

[6] To power this machine requires fuel; the complex sugar called glycogen is this fuel, and is stored in muscle fibers ready to provide the energy when sudden movement is required.

(for example, long-distance running), while those which move intensively (like in sprinting) need the fast fibers.

As one clear illustration of this, fish do not need to support their own weight, and so fish muscle is mainly composed of fast fibers, to allow rapid movement where necessary (to escape a predator, for example), and is much whiter than the meat of land animals as a result (unless they contain pigments that give a pink color, like salmon, or else are much larger and have naturally heavy and dense bodies that are not supported as much by the buoyancy of water and need to do more work to keep the fish floating, as for tuna and shark).

When an animal dies or is slaughtered, the muscle cells live on for a short period, and biochemical reactions continue to take place, including the consumption of glycogen. This produces lactic acid, which acidifies the meat (reducing its pH), helping to preserve it by inhibiting the action of both enzymes and microorganisms that would otherwise undesirably change the characteristics of the meat, and also resulting in the exudation of some water, giving meat a moist appearance. If death occurs in a stressed state, the animal will use up its glycogen stores early, and these post mortem changes do not occur, or occur to a lesser extent, and a darker, drier type of meat results; this informs the practices at point of conversion of animal to meat.

Within a few hours of death, rigor mortis sets in, as the glycogen runs out, and the muscles clench and adopt a rigid structure; controlling the processes before and around this time point are critical to getting a meat of desirable quality and texture.

A further key protein in meat is the aforementioned myoglobin, which has the function in the live animal of storing oxygen; such oxygen is delivered to it by its arguably better-known cousin, hemoglobin, which transports oxygen taken in through breathing through the blood to the tissues, where it hands it over to myoglobin to hold it in place until needed. Myoglobin and hemoglobin reflect their status in terms of whether they are currently holding oxygen or not by changing color, a bit like taxi lights. We know that arterial blood is bright red as it carries oxygen away from the heart, whereas oxygen-depleted blood on the way back from the bodily extremities through veins is blue. By comparison, myoglobin that has oxygen bound is red, while the depleted version is purplish-blue.[7] So, both the oxygen content and the level of myoglobin affect the color of meat, and tissues that need lots of energy (and hence oxygen) are darker because of its high levels than those that are more inert, like breast versus wing of chicken, for example.

Storing meat after slaughter allows enzymes to act on the proteins, breaking them down to reduce toughness after cooking. Sometimes, as we will see later in Chapter 17, we help this along by adding marinades to chemically transform the meat into a more palatable form, and sometimes we allow this to happen more naturally, as when we use dry aging. Sometimes, brute force is used to increase

---

[7] As we will see later in Chapter 14, much of the science of packaging meat involves taking steps to make sure myoglobin has enough oxygen to give it an appealing color.

the tenderness of meat, as in so-called mechanical tenderization, when blades, nee-
dles, or hammers are used to physically rupture muscle fibers and soften the meat.
However, this can transfer bacteria from the surface of the meat to the core, where
they might be harder to kill during cooking, and so this technique has been the
source of some food safety concerns.

Cooking and the transformation of meat proteins

In terms of understanding what happens when we cook meat, one of the most important proteins is the connective tissue protein called collagen, which is found in
particular in skin, bones, and the tendons accompanying muscles, as well as all parts
of the animal we consume as meat. Collagen is important because, when heated in
the presence of water, it denatures and changes into a new protein called gelatin,
which turns tough muscular structures into much more tender and palatable ones.
The differences among many cooking techniques then come down to differences in
how quickly and to what extent they achieve the conversion of collagen to gelatin.

Meat also contains fat as a further source of energy, fat in fact being a much
better energy storehouse, gram per gram, than carbohydrate. Fat is typically concentrated in particular seams within tissue, which can be trimmed off the outside
of meat pieces relatively easily. However, much finer seams of fat can run through
meat, called marbling, and play a key supporting role in ensuring tenderness, juiciness, and flavor.

So, when we cook a piece of meat, we are trying to balance several things, including the development of texture, the assurance of safety through killing of bacteria that might be lurking within, and the creation of flavors. As with any food
product, the core purpose has to be the assurance of safety, which sets the minimum
level of heating that is necessary, while individual tastes will then determine how
far a cook should go above that minimum baseline. The scale we can thus imagine
of cooking versus the characteristics of the cooked meat is determined by temperature, and we can equate different levels of "cookedness" with the temperature
reached, and hence the reactions caused at that temperature. We must also recognize that, within a piece of meat, depending on thickness, the surface will always
reach any given temperature ahead of the innermost places, and for the core to hit
a particular temperature means accepting that the parts nearer the heat source will
have attained notably higher levels of heating, and so will have changed more.

When we heat meat, the first changes will take place around 40 °C, at which the
proteins (specifically myosin) first start to denature and unfold (interestingly this
is not too far above the natural body temperature of the animal), while hitting 50
°C takes us into much more entangling reactions, and at around 60 or 70 °C the
collagen starts to soften by transformation into gelatin. The lattermost reaction is
critical for making tough meat more palatable, and so the more collagen-rich connective tissue is present the more the meat will need to be heated to soften it. On
the one hand, very sinewy meat with lots of connective tissue will soften best with

very long cooking at lower temperatures in the presence of lots of water, which is why such meat is best stewed, where the moist heat has time to melt the collagen into gelatin. Meat with low collagen levels, on the other hand, is fine for cooking by other means, such as roasting, where water is not needed to solubilize the collagen into a more digestible form.

At 60–70 °C, myoglobin loses its ability to bind oxygen and its original structure and starts to form complexes with other proteins. Around this temperature, a pink color appears, while around 10 °C later cell walls start to break down, and a gray color appears.

In terms of flavor development, some of the key reactions are the Maillard reactions, the reactions of certain types of sugars with certain side groups of proteins, and that give rise to changes in color, flavor, and aroma. In fact, much of what we regard as "meaty" flavor arises from these reactions and, as they happen at a rate that is highly dependent on temperature, we find them more in meat that has achieved the highest temperatures (over 100 °C) during cooking. The last actor in the meat protein transformation, actin, gives up its structure at around 70 °C, and this reduces juiciness and increases the toughness of the meat. Generally, meat is cooked so as to avoid excessive denaturation of this protein.

Typically, when cooking a piece of meat we want to ensure that the outside of the meat hits a temperature sufficient to brown it, and then the level to which we heat the interior depends on the level of connective tissue (collagen) and the desired end result in terms of "doneness." To achieve what is generally regarded as "medium-rare," a steak must be cooked to a temperature of around 55–60 °C, while "medium" and "well-done" require temperatures of 60–65 and around 75 °C, respectively. Of course, the exact interpretation of what constitutes "well-done" differs between countries, and I have learned that what in some continental European countries would be regarded as "well-done" might be in Ireland and some other countries be called "still running around the field."

We also need to remember that, as we heat the meat, the water that forms a major component will evaporate, and so the system becomes drier, which also forces proteins and other molecules into ever-closer proximity, encouraging reactions and interactions. When we cook bacon, for example, the loss of water concentrates flavors and promotes browning reactions and the development of particular flavor, aroma, and color molecules. Eventually, the molecules are forced into such close proximity that a sort of glassy state results, and the bacon becomes crunchy. Overcooking will go beyond a desirable state of crunchiness to a less-desirable (depending on taste) brittle crumbliness.

In the case of steaks, we want to use tender meat that is relatively free of connective tissue, and then cook to a brown surface with an interior tailored to the tastes of the eater. The difference among rare, medium, and well-done is simply a difference in the extent of denaturation, Maillard reactions, cellular breakdown, loss of oxygen and water, and a zillion other chemical phenomena. Instructing a waiter exactly how you like your steak cooked is simply a snappier summary of the extent of molecular havoc you desire to have wreaked on the piece of raw meat concerned.

In molecular terms, the more well-done the steak, the better the protein structure and initial chemistry of the meat will have been undone.

In Figure 4.1, we can see chicken before and after cooking. The process of cooking has transformed the pale soft pinkish-white meat into something far more firm and golden (thanks to the Maillard reaction), with an enticing sheen of juiciness. Legs, wings, and breast, with different functions in the live bird, will have yielded quite different cooked appearances and textures once the skin is peeled back, because of their different protein profiles, as discussed in the second section of this chapter.

Peer much more closely, however, using the electron microscope, and we can see the changes that have been wrought at a protein level. In both the raw and cooked versions of chicken meat seen in Figure 4.2, the fibrous nature of chicken meat is clearly evident, with loose strands of collagen-rich connective tissue like cobwebs on the outside of the fibers of actin and myosin. On cooking, the key difference is the initial shrinkage of the collagen coat, followed by its conversion to gelatin (and the much greater visibility of the muscle fibers), which makes the meat far more tender and easy to chew. The proteins have gradually contracted and exuded liquid, which makes the meat more tender, and so the texture of the cooked chicken has transformed into a much more palatable form, with protein chemistry driving this, while further changes that are due to heat-induced reactions have in parallel created the aroma and flavor we expect from cooked chicken.

Proteins in eggs

Another food product that shows off the critical role of proteins in determining food structure is the egg. Eggs contain two distinct materials or regions, the yolk and the white, with the yolk being a concentrated structure rich in proteins, lipids (including cholesterol), and vitamins, which is kept separate from the surrounding thick viscous liquid (egg white) by a membrane, like a little golden sack of nutrients. Differences in the types and levels of proteins present in each are critical to the difference between the different kinds of eggs; what they have in common is the fact that they denature on the application of heat.

Figure 4.3 shows the magnificent golden globule that is the yolk inside a cracked shell, while Figure 4.4 shows the yolk extracted in the preparation of dough. If we apply our powerful electron microscope to the egg we can start to see what a structural marvel it is, as shown in Figure 4.5. The shell is seen to be a crystalline calcium carbonate structure, covered on the inside with a delicate spiderweb of protein strands with lumps of the calcium carbonate within, looking for all the world like something within which we might expect to find an enwrapped insect or hobbit trapped.[8] This shell is actually quite porous, to allow the developing chick to breathe, while being strong enough to physically protect it.

---

[8] Captured by shellob, of course.

As just mentioned, the yolk and white parts contain different mixtures of proteins, which give very different cooked structures, and these can readily be seen in the lower two panels of Figure 4.5. When we imagine the consistency of cooked egg yolk, it can feel somewhat powdery and granular, and so it is not surprising that under our microscope it looks decidedly particulate, but in a highly regular and almost sculpted way. These large structures are called granules and contain cholesterol and protein. The cholesterol (specifically the nutritionally less-favored low-density lipoprotein [LDL] type) actually plays a key role in controlling the structure in both the raw and cooked egg. In strong contrast, the cooked white shows a much more random coagulated structure, which may account for the lack of granular texture when we eat it.

In chemistry terms, the difference between a soft egg (in terms of the yolk) and a hard egg comes down to the fact that the proteins in the two parts denature at different temperatures, with the white proteins being more susceptible to heat (denaturing around 63 °C) than those in the yolk (which need to get to around 70 °C). In the case of an egg, the principal outcome of denaturation is coagulation into a solid, and so heating below a critical temperature (at which the yolk will start to solidify) can give us a cooked, solid, outer white zone with a golden, creamy, runny yolk in the middle, while heating a little longer will bring the yolk to the point at which it finally gives in and starts to match the texture of its surroundings. This applies whether we boil, fry, or poach the egg, but if we mix the proteins by scrambling we can get coagulation under conditions that are essentially a mixture of the requirements of the individual parts.

Boiling an egg is actually a complex process of physics, despite its clichéd depiction as perhaps the simplest form of cooking that a person could manage to master. This is because the transformation of raw contents into the state of "cookedness" that fits individual tastes is highly dependent on a very variable raw material, as from egg to egg there will be differences in things like the exact shape and thickness of the shell and the proportion of yellow (yolk) and white parts, and so (as for so many food products) applying exactly the same conditions each and every time can lead to quite different end results. This is even before we consider the effect of duration of heating, which directly affects the level of denaturation and coagulation of the egg proteins, and hence the solidity or runniness of the egg. We will return to the complexity of heating in Chapter 10 when we consider the physics of cooking, but for now our interest is in the proteins.

When we cook an egg, there are two different reactions that take place, and these are quite separate, as we could take apart the white and yellow parts and cook them quite independently and find very different outcomes.

So, let's imagine a transparent egg with a temperature probe inside it, which we place carefully into a pot of water on a hot plate, which we start to heat. As the probe heads towards 60 °C, nothing happens, but once it crosses this limit things get interesting, and the clear thick liquid surrounding the golden globule of liquid yellow sunshine suspended within starts to be become turbid and turns white, with all apparent fluidity and movement ceasing at this point as liquid becomes solid. If we take the egg out of the water at this point, the white will be solid but the yolk liquid, and we have a soft-boiled egg. Leave it in the water, however, and we can watch the temperature climb further upward, until 70 °C is reached, and then our yellow liquid becomes solid, and our egg is hard-boiled. Different structures of the egg proteins, and different extents and strengths of bonds holding them in their "as-made-by-nature" native state, give the yolk proteins just that little bit more strength and resistance to heat than their surrounding cousins in the white, and allow a range of relative textures to be reached to meet the preferences of the consumer.

The key catalyst for the transformation is thus heat, which in this case we are applying to both make the consistency more palatable than the gooiness of a raw egg and to ensure safety, by killing any bacteria present in the raw material. These dual goals of heat, to cook and to render safe, as we will see, are at the heart of most food-heating processes. Cooking an egg is a complex proposition though, as we need the heat to gradually penetrate the shell and then transfer through the interior gradually so that it eventually heats the center. If we had in our hypothetical model several temperature sensors at different positions between shell wall and whatever might be judged to be the exact geometric center of the egg, we would see a distinct gradient of temperature, as the heat gradually worked its way through to the core. As we will also see, heat transfers faster in liquids than in solids, so a further complication is that the solidification of the white effectively insulates the yolk in that critical window between 63 and 70 °C, and the final coagulation of the yolk takes place in a kind of molecular slow motion, as the proteins therein slowly unfold and greatly expand, becoming entangled in and bonded to their neighbors, so that they enmesh the liquid water within the yolk in a tangled mass of unfolded cross-linked proteins, and it solidifies.[9]

In addition, the unfolding of these proteins exposes parts of these molecules that have specific properties that remain unexpressed when the proteins are in their native state. For example, exposure and breakdown of sulfur-containing amino acids lead to the distinctive, less-than-pleasant smell of overcooked eggs.

Finally, as we remove the heat, and surround the egg with cold water or air, the driving force behind the movement of heat reverses, and the heat starts to leave the egg for the cooler exterior, but does so slowly, such that on removal from the heat it continues to cook for a while yet, and gradually gets firmer with time as a result.

If we dilute the proteins in egg, by adding liquid before we cook a mixture of egg whites and yolks, we will still get coagulation on heating, but the proteins have been diluted and can't entangle so closely, and the consistency of the cooked mass becomes progressively more liquid than solid.

---

[9] In a further complication, the yolk is not stationary during cooking, but rather moves upward by gravity because it contains lipids, which we will encounter in Chapter 6. It is enough for us to note that these lipids make the yolk less dense, and so it rises when an egg is stationary. To be more precise, the surrounding white liquid is more dense, and so gravity drags it downward, and it pushes the yolk upward and out of its way.

The curious case of the tingling tongue and the overly wobbly jelly

Before leaving our discussion of proteins and their roles in food structure and the way in which enzymes can either help build up a structure (like chymosin in cheese) or else tear it down (as in meat), I would like to mention the aggressive enzymatic beast that lurks within pineapple, an enzyme called bromelain. This is found in the core of the fruit mostly, but can often be detected in fresh pineapple when we notice a tingling sensation in our mouths as we eat.

Would it be somewhat scary to mention that this is due to the aforementioned enzyme getting hungry and taking a snack on the proteins lining the inside of your mouth? What delicious (sweet yet tart) irony that, as you eat it, it refuses to go quietly and tries to eat you on the way. Luckily, this effect is rapidly reversed by growth of fresh cells, and the conditions lower down our digestive tract rapidly disarm this most aggressive ingestion.

This enzyme is also responsible for the frequent observation that fresh pineapple is not a good ingredient for a simple jelly, as the gel won't set properly. The main ingredient giving structure to traditional jelly is gelatin, the protein from meat, and the pineappular predator that is papain, on finding itself in its company, sees a snack and sets to work. In fact, bromelain can be added to steak to tenderize it precisely because of this ability to break down proteins.

Why are such problems not seen for tinned pineapple though? This is because, as we will see in Chapter 11, that tin has been subjected to extremely high temperatures, which even this tough fighter of an enzyme cannot resist, and it has been inactivated, leaving the fruit far more stable.

The secret codes hidden in food proteins

There are few cooler words, in my view, than "encryption," with its connotations of code-breaking, state secrets, keeping your credit card details safe, and perhaps someone suspended from a ventilation shaft while trying to hack into a top-secret computer. What does such a high-tech term have to do with food, and proteins in particular?

Encryption suggests the hiding of information in such a way that it is difficult or impossible to extract unless someone knows a method by which the code can be broken. This might involve scrambling words or information within a much larger set of letters and words, like 15 words out of 100,000 in a book that, when identified and read together, reveal a valuable secret.

When applied to proteins, the idea is that certain fragments of a protein might have properties different than, and indeed much more interesting than, the protein from which they came.

For example, casein and whey proteins in milk are valuable for nutritional purposes, and we have seen already how they are also critical for the properties and texture of dairy products. We have also seen how the casein proteins are broken down extensively during the ripening of cheese and that this modifies both the texture and flavor of that product, with tiny differences in the exact pattern of breakdown being part of what gives varietal characteristics.

When we consume any food protein, a similar process of breaking down (proteolysis) the protein molecules takes place because of enzymes such as trypsin and pepsin in our digestive tracts. This is how the proteins are converted to amino acids needed by our bodies. Some proteins, because of their structures, are very easy to digest by enzymes, while some are more resistant. In milk, the caseins have a relatively simple structure and break down easily, while the whey proteins have a more tightly folded compact structure that is harder for enzymes to penetrate and break down.[10]

However, in between full proteins and amino acids, the process of breakdown, in cheese or in our bodies, might give rise to a peptide of a few amino acids joined together that has a property of biological significance in our bodies. Such fragments are called bioactive peptides, because they are biologically active.

Examples of the effects that have been proposed for such peptides include reducing blood pressure, modifying our mood, combating the purportedly negative consequences of oxidation in our bodies, or helping kill harmful bacteria.

Enough information on activities for different peptides has now accumulated that databases exist of particular amino acid sequences that, if present in a peptide, exert particular effects, presumably by interacting with some site, receptors, or organs in our body and inducing them to do something different such as proves beneficial. In the case of the blood-pressure-modifying peptides, these act by inhibiting the activity of an enzyme in our bodies called the angiotensin-converting enzyme (ACE), and are for this reason called "ACE inhibitors." The function of ACE relates to production of an enzyme that increases blood pressure and can as a result increase hypertension and cardiac problems (it really is the ACE of hearts), so reducing the action of ACE is believed to be a health-promoting effect.

Lots of peptides that are known to inhibit ACE have been identified, and to produce a preparation that could make a particular food product or formulation have a heart-protective effect it is possible to tailor-make these peptides, if a suitable starting protein is identified.

To see how this works, imagine a hypothetical protein made of 26 amino acids in a chain, which we will say can be labeled, starting at one end of the molecule, as A, B, C, D, and so forth, up to Z at the far end of the chain[11]. Then we might spot that a particular peptide (which we might code-name "water") that has the sequence

---

[10] Caseins can be depicted as long straight molecules, and whey proteins as tightly balled-up spherical ones. Caseins are easy to digest; whey proteins are not. Caseins are like chips; whey proteins are like Brussels sprouts.

[11] In reality, amino acids are sometimes described by single-letter codes but, as there are only 20 commonly encountered amino acids, not all the letters of the alphabet are used.

HIJKLMNO is stuck in the middle of the chain (it is encrypted in the sequence, like a piece of information encoded in a longer string of information). To make more "water," we then just need to find an enzyme, or pair of enzymes, that likes to chop proteins with their scissors-like action between amino acids G and H and amino acids O and P (but not anywhere within the HIJKLMNO sequence). Adding this enzyme or enzymes to our protein then neatly pops out our peptide, and it is usually possible to design a method to separate it from the unwanted (but hopefully still useful for something) bits A to G and H to Z (perhaps using chromatography as discussed in the Appendix). Now we have a purified bioactive peptide that can be added to a product and, should the evidence of activity be strong enough to convince the relevant regulatory authorities, a label to inform consumers of its heart-healthiness can be added.

An example of a food product that already contains bioactive peptides is almost certainly cheese, as the protein in cheese is broken down so much that hundreds of peptides, several of which are likely to have biological impact, are present at the point of consumption. Centuries of successful consumption of mature cheese led to an optimistic conclusion that such activities, if present, are beneficial, but occasionally it has been suggested that milk-protein-derived peptides in cheese have darker effects. For example, it has been suggested that some peptides in cheese derived from beta-casein, called casomorphins, could have opioid-like properties, and make fast food products like cheeseburgers addictive.[12] Peptides are also starting to be added to foods, sometimes purified or sometimes in mixtures (like collagen peptides, currently in vogue for addition to coffee, ice cream and other products, because of supposedly beauty-boosting properties), while certain peptides from milk proteins are being added to oral health products because of their beneficial ability to strengthen teeth by transporting calcium.

Overall, proteins serve many functions in our diet, from texture to supply of essential amino acids, but it is clear that some food products contain peptide fragments that, whether released during manufacture or digestion or added deliberately, can influence processes in our bodies to modulate our health. Of course, it is likely that this has happened for as long as we have consumed protein-containing food and is part of the overall influence of food on our bodies, whether positive or negative, but as in so many other areas of food science and nutrition a rapidly increasing body research is enriching our understanding of this area, to allow such effects to hopefully be a part of the development of dietary recommendations and formulation and processing developments of the future.

Sticking proteins together: A literal approach

As we have discussed extensively, many protein-related transformations of critical importance to food result from proteins interacting to build larger structures, while

---

[12] This suggestion has appeared in, among other places, an interview in the 2004 documentary *Supersize Me* (an independent film by Morgan Spurlock).

others go in the opposite direction, with proteins breaking down to soften textures and provide flavor.

Sticking to the building-up side of the equation, another approach that literally involves building structures by making proteins work together involves adding an enzyme called transglutaminase. This enzyme is naturally found in our bodies, where it plays a key role in forming blood clots and wound healing.

Transglutaminase acts by creating new chemical bonds between amino acids, specifically two called glutamine and lysine. If these amino acids are on different protein molecules, the formation of such bonds leads to a strong new linkage between them. Transglutaminase is effectively protein glue.

In a food context, transglutaminase (derived from bacteria, rather than from blood) has been shown to have applications for things like making yogurt thicker and improving the texture of gluten-free bread, but it has probably attracted most attention for possible use in meat. It can literally take different pieces of meat, and even different types of meat (or strange meat–fish hybrids), and stick them together to create new textures, flavors, and even color contrast. Casing-free sausages can be made by using the enzyme to stick together meat pieces, for example, as can noodles made from meat and parcels in which different fillings are surrounded by sealed layers of meat. Because of the huge range of possibilities for innovation using transglutaminase, it is being seen increasingly as a tool in a range of creative high-end kitchens and is even sold under the refreshingly pragmatic name "meat glue."

Happily, when we eat any products in which transglutaminase has been used, the enzyme will often have been rendered inactive by cooking, but if not is readily digested, so it creates no risks to our own insides, while the new hybrid proteins created have not been found to have any associated health risks.

In conclusion, proteins confer the basic structures of many food raw materials, and many of the changes that take place when we cook or otherwise process foods relate to chemical changes in the structures of these proteins. We will come back to proteins later in this book, as when we discuss the structure of bread and the critical if sometimes allergenically problematic role of a complex called gluten, but for now we move on to what, for many foods, are the other major components, which also relate to and influence structure, which are carbohydrates and fats.

5

From Sweetness to Structure
CARBOHYDRATES IN FOOD

The sweetest things: Sugar in food

When we refer to food as containing "sugar," we tend to picture white crystals we buy in bags or pour from sachets into our coffee, but to a food scientist sugar is not a single thing, but a type of thing. While, most commonly, when we say sugar we refer to sucrose, in reality there are many sugars that can be found in (or added to) food.

They all have in common the chemical characteristic that they are carbohydrates, which means, as the name suggests, that they are based on carbon and water (giving hydrated carbon), and indeed the three core elements found in all sugars are carbon, hydrogen, and oxygen.

One of the simplest sugars, molecularly speaking, is glucose, in which there are 6 carbon atoms, 6 oxygen atoms and 12 hydrogen atoms (so it is like 6 carbons plus 6 water molecules, as each water molecule has two hydrogens to one oxygen). These are arranged, not in a long chain, as in the proteins we discussed in the previous chapter, but rather in a ring structure (not technically round, but more like a hexagon might appear if you gave it a good twist). Glucose is the main sugar found in biology, being found in our bodies and also produced by plants by photosynthesis.

Another simple sugar is fructose, found widely in fruit and honey, which has an even simpler structure, a pentagon structure, but again has 6 carbon atoms bound together with 6 oxygens and 12 hydrogens, while a rarer one (at least in its unbound state) is galactose, again with the same number of the core atoms but arranged in yet another slightly different state.

Here is a wonderful example of the significance of chemistry for food, where the exact same number of atoms of the same three elements can naturally be found in (at least) three different arrangements, which reflect subtly different molecular shapes but yet give compounds that differ greatly in their sweetness, solubility, reaction with other components in food, and many other properties.

These three sugars mentioned all have one core ring to their structure and cannot be subdivided into simpler molecules that fit the definition of a sugar. As the simplest sugar units, they are for this reason referred to as monosaccharides (as they have one indivisible sugar or saccharide structure—wonky ring—in each molecule).

The sugars with the next level of complexity are called disaccharides, which as can be guessed are made from two monosaccharides being bonded together. What we encounter most often in bags labeled "sugar," and what is envisaged when this term is used is more accurately called sucrose. This is a disaccharide, each molecule of which includes a molecule of glucose and one of fructose chemically bonded together. The sugar found in milk, lactose, is also a disaccharide, in this case formed by the joining together in chemical bonding of one molecule each of glucose and galactose.

The bonds that join disaccharides together can under certain circumstances be broken, in a sort of molecular divorce, to give the independent monosaccharides. In the case of lactose, this can happen when we add an enzyme called beta-galactosidase. This enzyme can be bought in pharmacies under the name of "lactase" and added in drops to milk or infant formula in cases of lactose intolerance, when the joined-up molecule causes discomfort that the simpler individual sugars do not. Most mammals, including humans, are born with the ability to produce this enzyme naturally, but those who suffer lactose intolerance lose this ability in later life (this includes large proportions of the populations of Africa and Asia).

As the sugars we have discussed to date all include the same number of the same atoms, just arranged differently, it is also perhaps not surprising that they can interconvert into each other under certain circumstances. For example, when milk is heated to very high temperatures (over 100 °C), lactose can convert to lactulose, in which the glucose part has transformed into fructose and the galactose remains as it was but is now attached to a new partner.

Why are sugars important in food?

Sugars in food have many functions, some familiar and some perhaps less so. The first and most obvious is conferring flavor and tingling those receptors on our tongues that are sensitive to sweetness. Not all sugars are equally sweet though. For example, on the one hand, lactose is significantly less sweet than sucrose and, if milk contained the level of sucrose that it does lactose, it would probably be much more popular with children! On the other hand, a side effect of the separation of lactose into glucose and galactose, as mentioned earlier, is that the reduced-lactose milk is sweeter, because glucose is sweeter than lactose.

Sugar also has perhaps a less-well-recognized role in the preservation of food, a property that has been very successfully applied in jams for centuries or longer and is one of the forms of traditional chemical preservation that we will discuss in later chapters.

Artificial sweeteners are of interest as alternatives to sugars because they supply the sweetness, by stimulating the sweet taste receptors, without the baggage of all the calories associated with conventional sugars. Their discovery allowed the development of sweet foods and beverages that were sweet but less calorific than if they contained regular sugars. Interestingly, several artificial sweeteners have been discovered by accident, such as saccharin (in 1878) and aspartame (in 1965). Saccharin was discovered in the process of a researcher's doing work on impurities in natural sugar, while aspartame was discovered as part of work on an anti-ulcer drug—in both cases the discovery story features a "eureka" moment where the discoverers licked their fingers and got a surprising burst of sweetness, followed by a retracing of steps to identify the source in an unlikely lab vessel.

Building upward: Polysaccharides in food

I have so far described monosaccharides and disaccharides, which contain one or two simple sugars, respectively, per molecule, but many more complex molecules found in food are dependent on simple sugars as building blocks.

These include trisaccharides (not that common in food), tetrasaccharides, and larger assemblies of still reasonably small numbers of sugars called oligosaccharides.[1] One common place to encounter these molecules is human milk, where there are quite significant levels of them compared, for example, with bovine milk, and their function is thus thought to be biologically significant.

Go beyond "oligo" though, and we are into the realm of the poly.[2] Polysaccharides are large molecules composed of hundreds or thousands of simple sugar molecules assembled into chains or other complex structures, and these polysaccharides are hugely important in food, from bread to jam to fake caviar, as we will see. When we talk of starches, gums, and stabilizers in food, we are in each case using common terms for polysaccharides.[3]

Polysaccharides are found commonly throughout nature (which is how they end up in food), and common sources of them include seaweed (perhaps surprisingly, the source of many food ingredients, like agar, alginate, and carrageenan), cereals and potatoes (which store energy in the form of starch), fruit (in which pectin gives structure), and even microbes (xanthan gum, an increasingly common commodity in many kitchens, is produced by a helpful tame bacterium).

As with so many molecules in food, tiny differences in chemistry can have a massive impact on the properties of polysaccharides. For example, plants contain

---

[1] This is a bit like a chemical analogy of how Terry Pratchett said trolls counted, which was "one, two, three, many."

[2] Perhaps a land called Polynesia? Molecules that comprise large numbers of simpler building blocks, whether in food or plastics, are called polymers.

[3] The renaming is perhaps understandable, as we could never picture kids asking for a bag of "wine polysaccharides," for example.

starch and cellulose, but one is a food material while the other is the plant equivalent of bricks that build structures like decidedly un-foody wood. We can digest
starch readily because it is intended to store energy when not needed and release
it when we need the power of its constituent sugars, while cellulose is almost im-
pervious to breakdown. While both comprise structures built from simple glucose
molecules, the key difference comes down to how these are organized and bound to
each other. In cellulose, the packing of rings of glucose molecules is so tight that
they cannot readily be pulled apart to break them down, while starch has a relatively looser and more open structure, which is easier to digest.

Starch itself is actually composed of two different polymers of glucose, called
amylose and amylopectin. In the case of amylose, the glucose molecules are arranged in long chains, while in amylopectin some of these chains have the molecular equivalent of arms and legs, with branches of glucose molecules off long,
straight, central chains. An amylose molecule might be imagined like the letter "l,"
while the structure of amylopectin might more resemble a "k" or a "y," or even a
Christmas tree–like array of branches. Amylopectin molecules are also much bigger
than those of amylose. Starches that contain higher levels of amylopectin are often
referred to as "waxy" starches.

The proportions of these two molecules influence the cooking properties of a
material, with those that are high in amylose, like long-grained rice, requiring more
water and higher temperatures to cook than those that are high in amylopectin, like
short-grained rice.

Starches can also be chemically modified to influence their gelling and thickening
properties, or broken down with enzymes (hydrolyzed) to release large amounts
of their constituent simple sugars. High-fructose corn syrup is produced by the
breaking down of cornstarch by use of acid or enzymes, followed by the transformation of some of the glucose to fructose, yielding a low-cost (but sometimes controversial, because of both the potential to increase calorific intake and concerns
about possible contaminants) sweetener for food and beverages.

One common type of modified starch is maltodextrin, which is produced from
amylose molecules by breaking down the long chains of glucose molecules into
much shorter stretches (fewer than 20 glucose units long). Maltodextrins are quite
sweet, very soluble in water (making them good thickeners), and are commonly
added to snacks as a filler because of their ability to bind to fats, thereby reducing
greasiness while resisting absorption of water during storage.

In Figure 5.1A, we can see a single potato, while Figure 5.1B could reasonably be
guessed to contain a number of such floury wonders. However, Figure 5.1B actually
shows the starch granules within the potato, where the key attributes of "flouriness"
that give the potato its multitudinous applications reside. But how does this happen
when we cook these dull-looking blobs (both at the visible and the invisible scale)?

To understand why polysaccharides are so important in potatoes, and in food
in general, we need to understand how they work, and in particular how they help
to control the movement of water; to do this we have to consider something called

rheology, which is basically the science of the structural properties of food that polysaccharides help to determine.

Rheology and food texture

When we open a can or other container and pour out its liquid contents, what happens next depends on what is called rheology, or the science of how materials flow when subjected to a force.

On the one hand, for example, the simplest rheological behavior is that of water, which flows readily under most circumstances. Unless we go to very low or very high temperatures, where freezing or boiling starts to happen, its behavior is pretty much unaffected by temperature. On the other hand, a material like tomato ketchup behaves very differently under different circumstances, frequently needing a firm shake to induce it to flow, whereupon it usually flows in thick, sluggish manner. Also, we can picture starchy materials like powdered soup being stirred in water in a saucepan, and changing their behavior quite markedly as we heat them, often suddenly seeming to thicken once it starts to get hot.

What causes such changes and how can they be measured, understood, and then controlled?

Let's take that starchy soup first, and consider the behavior of the starch grains that are the key controllers of flow and texture in that product. In the uncooked state, the starch is tightly packed into what are called granules, which is their home in the plant materials where starch is naturally found. Each granule contains a large number of complex polysaccharide molecules, packed extremely efficiently and tightly. Picture them as clothes neatly and efficiently folded, taking up every spare bit of space in a suitcase, or as pristine sheets of paper in a pack. Now imagine taking these clothes or sheets of paper out and crumpling them up, and then try and repack them. Will they fit back in their original package as effectively as they did, or will their new chaotic and disordered form take up far more space than they did before? Exactly.

The starch granule is the original suitcase or pack of paper, and when we cook them beyond a critical temperature (called the gelatinization temperature) the suitcase or paper package bursts open, their contents spill out, and the neatness and efficiency of packing are completely lost.

In the pre-gelatinized state, water will flow around the granules unimpeded, but once these burst open the starch molecules unfold their long space-filling molecules, and the clear flow path becomes entangled with these starch molecules, getting in the way of the water and each other; as in so many other examples in this book, a molecular-level change has resulted in a meal-level visible change—thickened soup!

To use another analogy, picture a group of ice dancers spinning really fast on skates, but with their arms tightly by their sides; they don't get in each other's way much, and a lot can fit in a certain-sized rink. Ask them to raise their arms as they spin, so that they all take up much more room and have now dangling, entangling appendages; chaos, or at best impaired flow, results. Bursting the starch granules extends the metaphorical arms, while the energy of heating causes the rapid spinning.

The behavior of starch in water also relates to the relative solubility of the two main constituents, amylose and amylopectin, as the former is soluble in hot water, while the latter isn't very water soluble at all. So, when we heat starch in water, the amylose leaches out into the water and leaves spaces that water can fill, which contributes to the swelling that ensues. If we don't disperse the starch well during the heating, lumps in which water is entrapped will form readily during this stage.

Starch is just one example of a polysaccharide that we use to modify the texture of food. A huge range of polysaccharides is available today for use as ingredients in food products or meals, and they come from a range of sources, including plant, marine, microbial, and synthetic origins. What most polysaccharides have in common is an affinity for water first and foremost, and the ability to soak up and thereby immobilize or otherwise restrict the free mobility of water.

To consider the power of these molecules to modify the behavior of water, consider an apple and an equivalent weight of milk—which contains more water? That seems easy, as milk is a liquid, but the fruit is clearly a solid (even if it is clearly juicy, suggesting a high content of water). However, the answer is that the apple contains well over 90% water, while milk has less than 90%, and the difference in structures is clearly not determined by the amount of water that is present, but rather its freedom to maneuver.

Milk is full of hydrophobic molecules, which actively repel water, and certainly don't soak it up or bind it, such as proteins and fats, as discussed in other chapters. So, the water that is present is "free" to act as a liquid, without anything cramping its style.

In the apple (or any fruit, really), the water has far more restrictions on its mobility, as the fruit is full of complex polysaccharides that have a massive affinity for water. These molecules entrap so much water that the structure of the fruit in essence acts as a massive sponge, and far more water can be contained per unit weight than might be expected to be possible.

The key polysaccharide in many fruits is pectin, which has the particular ability to bind water and form a gel in acidic environments, just as is found in most fruit products. This is why pectin is a common ingredient in homemade jams, where cooking it with the fruit causes it to soak up free liquid and form a perfectly jammy gel.

One perhaps surprising source of many polysaccharides that are commonly used as food ingredients is seaweed, which has yielded many viscosity-modifying polysaccharides[4] that are food ingredients, such as carrageenan, agar, and alginate.

---

[4] Ireland's beaches are full of seaweed, and visitors probably don't associate the stuff that gets in their way as they shiver in our Atlantic coastal waters with so many products they consume, including even the ice cream with which they may later cheer themselves up after their chilly dip. Many of these beaches are on what is today branded as the Wild Atlantic Way, a beautiful route around the

There are a number of different types of carrageenan that can be used to thicken liquids, some of which give thicker gels and some of which give softer, brittle gels. Carrageenans can be often found in chocolate milk, where they form a weak, almost imperceptible, gel in the liquid, which still has enough structure to entrap the large cocoa particles that are key to the product's characteristics and stop them from settling to the bottom of the bottle or container.

Agar is another frequently used thickener, a solution of which is liquid at warm temperatures but solidifies into a firm jelly when cooled, a property that has long been exploited in Asian cooking and now used more widely as a gelatin or egg replacer in jams, jellies, and desserts.

Alginate from seaweed has a very specific dependency on calcium for solidification, and, while the form of alginate bound to sodium can be completely liquid in water, adding calcium to it can result in an abrupt and rather startling transformation to a relatively rigid solid. This can be exploited in what is a neat culinary trick; a solution of sodium alginate is prepared and dropped slowly into a bath of water containing calcium, for example through an eyedropper or syringe. Each drop will fall from the nozzle, thanks to the force of gravity, as a perfect sphere. Once it enters the calcium bath, however, an immediate transformation takes place, and the outer surface solidifies, followed gradually by the interior as the calcium gradually percolates inward, and a perfectly spherical bead results. This can easily be exploited to make fruit beads to add to a dessert or yogurt by mixing fruit puree with the alginate and dropping that into the calcium bath. Alternatively, in a cool marine-themed trick, the sodium alginate can be mixed with squid ink or some other darkly colored material, and the beads that are scooped from the calcium bath create a reasonable analogue for the far more expensive caviar.

With care and practice, much larger structures can be created with alginate and calcium than beads, such as the cocktail of hazelnut milk and apple brandy that forms a smooth globule in Figure 5.2. Here, a cocktail of the apple brandy and hazelnut extract have been mixed with sodium alginate, which gives a thick but still liquidy liquid. When a blob of this liquid on its spoon was immersed gently in a bath of calcium chloride solution, the outside solidified rapidly as the sodium and calcium ions exchanged places, creating insoluble calcium alginate, and then the inside gradually gelled as calcium diffused inward. A quick wash later, after enough time had been judged to have elapsed to give a decent structure, the "solid cocktail" resulted.

Calcium and minerals like that of course are very important for our bodies and, along with vitamins, are hugely important in our bodies even if present at tiny levels. Mostly they are just carried by food into our bodies, but sometimes, as with alginate, they play a more functional role.

---

entire western half of the country's coastline; this surely creates future opportunities for branded food products, such as those made from Wild Atlantic Whey.

Another polysaccharide with some neat special powers that can be exploited in food applications is called xanthan gum, which is produced by a particular type of microorganism (called *Xanthomonas campestris*[5]). The power of xanthan gum can be readily demonstrated by looking at a solution of it in which some particles like colored beads have been dropped. At rest, the solution is extremely thick, and the beads will remain suspended in an apparently gravity-defying manner. Apply a little force, though, such as tilting the container, and the whole mixture flows; stop pouring, and the immobilization returns. A solution of xanthan gum behaves a little like a children's game where, while the music is playing, energetic motion such as dancing is freely permitted and even strongly encouraged but, once the music stops, the children have to pretend to be statues. The molecules of xanthan gum are the children; the force is the music.

At a molecular level, the xanthan gum molecules can be envisaged as rods, which when at rest like to arrange at right angles to each other, forming a lattice-like structure that entraps water. Apply a force, however, and the bonds between molecules in this structure are sundered, and the molecules move to a side-by-side conformation in which they can flow. A certain critical level of force is needed to kick off this process, but after that the bonds have broken and flow ensues, at a rate that is higher when more force has been applied. A material like this is referred to as being shear-thinning, where shear is the term used to refer to an applied force (particularly a pushing, stretching, or pulling force, as opposed to, for example, a mixing or pressing force), and the thinning bit means that the result of the shear is that the material being sheared becomes less viscous and thereby thinner. The design of the perfect ketchup consistency through careful addition of xanthan gum could be perhaps considered as shear genius.

This ability to move–stop–move is very useful indeed, which is why xanthan gum is found in lots of food systems. For example, consider a salad cream or dressing, which is poured over mixed green leaves. The objective is for these leaves to be coated, rather than the dressing flowing off and pooling at the boom of the bowl. In the presence of xanthan gum, the dressing pours fine, but when it meets a sudden deceleration, to the point of stillness, however fleeting, on hitting the solid leaf, liquidity vanishes and the leaves become flavorfully coated.

Flavor release from food can also be manipulated by molecules like xanthan gum, as thicker materials stay in the mouth longer and release flavors more slowly, because of the greater time for reaction with flavor receptors, and more time for volatile aroma compounds to be released and picked up (hopefully favorably, or even flavorably) by the ultrasensitive sensors in our noses.

This is one reason why it is today found in many kitchens, where it is used frequently in bakery applications, in particular recipes for gluten-free products, where it gives mixes a stickiness which is otherwise contributed by gluten.

---

[5] Not, apparently, despite what might be guessed, named after anything to do with Greek deities or Amazon warriors.

Other unusual polysaccharide properties that are exploited both in factory and occasionally in kitchens include the ability of methylcellulose to behave in the opposite direction to most thickeners and set as it gets hotter while melting as it gets cooler, rather than vice versa. This polysaccharide, made by taking the boring and indigestible cellulose found in plants and chemically modifying it (adding the methyl bit both in name and in structure), can be exploited in such concepts as hot ice cream, which melts as it cools, and melt-in-the-mouth noodles, which are actually being cooled on eating rather than the usual melting that takes place because of warming in the mouth. Liquids containing methylcellulose can also be squirted into hot dishes like soup, and the heat-induced change in the structure of the molecules results in the formation of noodles or other solid shapes. These unique properties make methylcellulose a favorite ingredient in the world of molecular gastronomy.

The Maillard reaction

In many food products, a characteristic change that takes place during exposure to high temperatures, as in heating, cooking, or baking, is the development of brown colors, and some products (especially those containing milk or milk-derived ingredients) are known to be better at this than others.

The key to such changes is the Maillard reaction, which is actually a very complicated set of reactions and pathways discovered by a French scientist called Louis-Camille Maillard.[6] The key to a good Maillard reaction is the reaction between certain types of sugars (of which lactose in milk is a good example) and proteins, plus significant heat (although, once kicked off, the color-development reactions can proceed quite happily at lower temperatures). The best proteins for Maillard reactions are those that contain the amino acid lysine; this amino acid is very common in milk proteins, and so milk is a material that is particularly susceptible to these reactions, containing as it does both the ideal sugar and a very suitable protein.

This can readily be seen in dairy products that have been heated to different extents: Compare pasteurized milk with ultra-high-temperature (UHT) milk with sterilized evaporated milk that has been heated in a can, and a gradient of burgeoning brownness can be clearly seen, going in that order, as the progressively more severe heat treatments encountered give stronger colors. The different samples have similarly different flavors, as many of the products of the Maillard reactions have strong flavor and aroma characteristics (from cooked to fruity, fresh, or downright nasty). Products with strong aromas and dark colors, from toast to beer to

---

[6] Who actually made the discovery while working on the reactions of sugars and proteins in cells and not food at all. Only after his death were the reactions he studied connected to the changes made on heating food, and he received posthumous credit for something he might be surprised to find made him famous.

coffee, all owe their characteristics to Maillard reactions caused during their processing or final cooking before consumption.

Indeed, the diversity of pathways kicked off under the umbrella of what Maillard first described is quite dazzling, and any textbook diagram of the Maillard reaction has so many arrows, pathways, and products as to resemble a map of an infeasibly complex electrical circuit created by an engineer having a very bad day. Figure 5.3 shows a simplified illustration of the possible routes by which sugars and proteins can interact or change in food when we apply heat.

Nonetheless, when simple molecules are kicked down one or more of these paths, the destination can define a food product, with the flavor of cooked meat, for example, being hugely dependent on the creation of specific compounds through Maillard reactions. The other key outcome of the Maillard reactions is of course the development of brown colors, which is due to a family of compounds called melanoidins, and the reactions are often referred to as nonenzymatic browning, to reflect this.

Sugars can also react to heat in quite dramatic ways even in the absence of proteins, and caramelization reactions, which we will discuss later, depend on specifically such reactions.

Microscopic icebergs and the titanic role of sugars in ice cream

Besides sweetness, sugars play several key roles in food, some of which might be quite unexpected. For example, in ice cream we think of sugar as making the product sweet, and that is of course one key function. However, there is a lot more to the role of sugar in ice cream than making it sweet (which is really quite a bonus).

To understand the primary function, think about what would happen if we freeze milk, say, in a domestic freezer in ice cube trays. Does it look like ice cream and, more important, does it have the same light structure? No, it looks like frozen milk and has a consistency like that of an ice cube.

Part of the difference is due to the fact that ice cream contains a lot of air bubbles to break up and lighten the structure (in other words, ice cream is a foam). The other key difference, though, relates to the presence of sugars and the fact that they are basically getting in the way of the water as it tries to organize its molecules into the neat, orderly, crystalline structure of ice. Picture a lot of people standing still while trying to arrange into a shape for a photo by holding hands and looking very neat and orderly, and then release some enthusiastic puppies into the group. The puppies keep interrupting things by jumping up on the participants and demanding petting, and preventing the neat and orderly lines from lining up. People here are the water molecules, sugar molecules the puppies (appropriate, given how sweet they both are!).

So, if we add sugar to water, it makes it a little harder to freeze and, whereas pure water might freeze at 0 °C, water with sugar in it needs to go to colder temperatures before this happens. The more sugar is dissolved, the lower the freezing point. The same happens incidentally with anything dissolved in water, which is why we add salt to roads and paths on a frosty morning. The salt dissolves in some of the free water around the ice and lowers the freezing point, which means that, at that cold morning temperature, it is harder for the ice to remain frozen, and more thawing occurs.

In fact, we could make ice cream from a mix in which we substituted salt for sugar, and salt is actually better than sugar for reducing the freezing point of water, if the two were present at similar levels. However, the resulting odd-flavored product would melt much more readily than the sugary version, even if a market for salty ice cream (beyond the current taste for salted caramel as a flavor for most sweet things) were ever identified.

In the case of freezing our sugary, milky, ice cream mix, let's say we need to go to a few degrees below zero for the mix to freeze, which eventually happens as we cool it using very low temperatures in an ice cream freezer. When freezing kicks in, what forms are some pure water ice crystals, which remove some water from the system, so the remaining sugar (which is outside these crystals) is now dissolved in a little less water. That reduced volume of water has a higher sugar concentration and a lower freezing point, and is even harder to freeze, and so on. So, as the mix is cooled to progressively colder temperatures, the mixture is effectively continuously becoming harder to freeze, and when the final freezing temperature is reached (maybe 18 °C below zero, the temperature of a typical domestic freezer), the ice cream mix has internally separated into a population of ice crystals surrounded by a thick sugar syrup that is too rich in sugar to even contemplate freezing at that temperature.

So ice cream is definitely not frozen milk, and is not even a solid. It is actually a very thick liquid (sugar solution) in which float, like slow-moving icebergs, ice crystals, which give apparent solidity but do not actually link up into a continuous frozen structure. Within this strange soup also float inclusions or other flavors we have added for additional pleasure, plus the entrapped bubbles of air that were also incorporated during freezing and whipping.

So, one of the key principles of making ice cream involves sugars controlling the freezing and initial texture of the product. Keeping it in this state, though, is a different challenge, and one that needs the input and special functionalities of some other carbohydrate-based ingredients, specifically polysaccharides.

I suggested earlier that sugar determines the basic frozen structure of ice cream, but the problem is that this doesn't always last very long. When storing ice cream in a freezer in our homes, the temperature is far from constant as, every time the freezer door opens and closes, to add new purchases or remove products for consumption or cooking, a little warm air enters. At this point, the temperature rises, even by a degree or so, and when it does some of our ice crystals in the ice cream melt (as more water will be liquid at –17 or –16 °C than at our ideal –18). Then, when the door closes, the mechanics of the freezer work to bring the temperature back to –18, and the ice refreezes. The problem though is how it refreezes or, to be even more precise, where.

For the ice cream to have the smooth cold but not icy structure we want, a key principle is for the ice crystals to be below a certain size at which we would detect the crystals as individual particles on our tongue. Above this critical size, the ice cream would taste "gritty" or "icy" on the tongue. If we control our ice cream manufacture well (particularly the rate of freezing and how well we mechanically disperse the crystals as they form), this is relatively easy to achieve but, when the initial crystals melt and refreeze during storage, the trend is very much toward small crystals melting, with the water migrating to the surface of other crystals and freezing there, making them larger. So, basically, the ice crystals grow bigger during storage, and the ice cream gets progressively grittier as a result.

This is why sometimes, when clearing out a freezer or icebox, people find forgotten abandoned samples of ice cream or other products at the bottom that have been there for years. When taken out, these look like something from a horror movie set at the South Pole, with large lumps of ice where nice food once was found and that effectively disintegrate on thawing. Tasting such products would give a new meaning to frostbite. This is the ultimate fate of all ice cream, to grow progressively icier and less lovely because of the ravages of ice growth; what we can control through careful formulation and ingredient selection is the rate at which this happens and how long the ice cream will remain at its optimal state of delectability.

This is where the polysaccharides on the ingredient list come in. As I have said, their main talent is immobilizing water and controlling its movement. In the case of the urge of ice crystals to grow and grow in ice cream, this becomes a critical gift. As the ice crystals melt, some of the water molecules become free from their ordered lattice prison and look to escape. The problem is that, when the next temperature dip finds them in the open, they tend to get absorbed onto existing other crystals that they have managed to get closer to while celebrating their newly found but tragically short-lived liquidity, and so these grow bigger and the ice cream grows grittier.

The polysaccharides (such as carrageenan or guar gum, which is extracted from a bean commonly found in India and Pakistan) build cages or walls around the ice crystals, which block the ready escape of the liquid water molecules. These are then trapped near their original crystal home and, when the temperature drops, they have no choice but to turn around glumly, trudge back home, and resume their icy station in the lattice to which they had thought they had waved their farewells. So, the original ice crystals just pulse gently with the temperature fluctuations, decreasing a little in size and then going back their original dimensions each time the freezer door opens and closes. The protective wall isn't perfect, alas, and some migration to larger crystals takes place (the escapees find gaps in the wall, perhaps, or loose bars in the molecular cage), and the ultimate depressing icy fate of the once-magnificent dessert still beckons, but this distressing end happens a lot more slowly than would happen without these molecules being present.

So, in summary, in ice cream we find small and large sugar molecules, some of which make the frozen structure on day one, while the others then report for duty to try and stop age from undoing their good work. They really do a lot more than making the product sweet and are the reason why many frozen dessert products, even the simplest frozen fruit "pops," include sugars or polysaccharides, or both, as they make the difference between a hard icy lump of a product and a nice smooth texture.

The size of ice crystals also influences the perceived coldness of the ice cream, as smaller ice crystals have a much greater surface area than large ones containing the same amount of frozen water, and small ones melt much faster as a result and absorb much more heat to fuel this transition (as discussed more in Chapter 11). So, very rapidly frozen ice cream (such as shot-style products with really small balls of ice cream), in which extremely small crystals have been formed by using liquid nitrogen during the freezing process to give a huge temperature difference and hence cooling force, feels much colder out of the freezer than a more conventionally frozen product.

The insane levels of coldness that can be achieved with liquid nitrogen can even be exploited in cryo-cooking, in which a quick immersion in liquid nitrogen freezes the outside of a piece of meat, for example, which is then fried to give a perfectly cooked outer surface without overheating the interior. The key principle here is an incredibly high temperature difference between the deep-deep-deep-frozen surface and the hot oil, which, as we will see in Chapter 11, results in extremely rapid heat transfer and cooking.

Honey and sweets

Another familiar example of a food product whose characteristics are dominated by the properties of sugars is honey. Sweet, syrupy, and prone to crystallization and solidification if underminded, honey is another foody manifestation of the chemistry of sugar. Honey is also an ancient food (being well established as a food in cultures as ancient as those of Egypt, Babylon, and Sumer[7]), resulting from a long-standing if rather one-sided agreement between humans and bees, whereby the smaller partner does the work, collecting nectar from flowers and converting it to simpler sugars that are loaded into the hive as honeycombs to feed their young[8]. This bee-based bioconversion to simpler sugars is nature's way of providing the greatest energy hit for the young, as the smaller molecules are more soluble, allowing more energy to be packed into a fixed volume of liquid; honey is bee espresso with an extra shot.

So, honey at harvest contains mostly the fruit sugar fructose, as well as some more complex sugars (dextrose), water, and low levels of gums, minerals, vitamins, and acids (which render it slightly acidic). It also contains some flavor compounds from the plant source, which tone down what otherwise would be an intensely

---

[7] It was also fermented by the Vikings into the alcoholic beverage called mead.

[8] A very common pairing in food and nonfood contexts is that of milk and honey, and one particular area where they share identity is that they are both products that humans take without permission from one species that which was originally created to feed the young of that species.

sweet flavor due to the high sugar content, and some further plant components that confer the golden color. In processing the honey for consumption by humans, some of the waxy components are removed, and some additional conversion of the larger sugars to simpler ones may be undertaken.

Honey is well regarded for many properties, including health promotion (because of the content of B vitamins) and brain stimulation (as attested by W. T. Pooh), but also as a preservative, due to the high level of sugar. This is an example of how sugars can act as both enablers of bacterial growth (as is the basis for most types of fermentation) and as inhibitors of this. As with many things in life, the difference between promotion and inhibition is simply a matter of level.

Honey is full of sugar, but is highly unlikely to spoil or show bacterial growth. Indeed, its protective properties in this regard have made it a long-used additive to other foods, as in placing a microbially repellent coating around the outside of meats, for example. The point at which consumers might determine that their bottle of honey is no longer acceptable (i.e., the end of shelf life) is likely to be when a thick whitish-grayish solid material has appeared in the once-clear golden liquid and the mass has started to solidify. The fact that heating the liquid can (at least temporarily) reverse this change attests to its sugary cause and is another example of a food "spoiling" by a means other than the microbiological.

Honey is typically close to its highest possible degree of saturation with sugars, which means that it cannot absorb any more dissolved molecules in the amount of water present, and anything else that is added simply cannot dissolve and will remain suspended (given the viscosity) or (perhaps eventually) settle out. Should a bacterial cell land from the air into an open pot of honey, it will find itself in a hostile environment in which sweetness is deceptively deadly. Each bacterial cell contains a large amount of water and is surrounded by a membrane that controls the movement of water and solids between the inside of the cell and its outside environment, which is required for normal operations and survival. However, in a sugar-rich environment, a stark contrast exists between the sugar and water balance inside and outside the cell, and this creates an irresistible natural force (called osmotic pressure) whereby nature tries to bring the two sides of the bacterium's membrane into balance by diluting down the outside environment's sugar level. Considering the somewhat inequal proportions of liquid that might exist inside a cell versus a surrounding pot of honey shows this to be a rather foolish and quixotic tendency, but logic doesn't trump physics, and the cell, against its better instincts, donates its vital inner liquids to the outside world, and in the process becomes dehydrated to the point of death.

So, spies and microbes alike can hence fall prey to the honey trap, and this ability of sufficiently high levels of sugar to kill bacteria is exploited in many food products, including the preservation of jam by the addition of high levels of sugar.

If we go to higher levels of sugar, we can go beyond the liquid character of honey to solid products, as there isn't enough water present to keep the sugar molecules sufficiently far apart that the product is still (barely) dominated by the properties of water, and will flow (if somewhat sluggishly). Go beyond these levels and the water loses control, and the sugar molecules take over the structure, interlocking into shapes with different structures and interactions that give us delights from the soft chewiness of toffee to the brittle glass of hard candies (sometimes called boiled sweets). The sugar molecules will be chemically attracted to each other, but the presence of water molecules in the way will prevent these from allowing natural bonds and structures to form; remove these and nothing stands in the way of a sugary solidification.

To make these products, water must be evaporated from a sugar solution to bring the sugar molecules that remain behind into ever-closer proximity and give us our solid structure. This is a complex process, though, as the presence of sugar increases the boiling point of water, with a direct relationship between sugar content and boiling point. For this reason, as we remove water, the sugar becomes more concentrated in a smaller amount of remaining liquid and the mixture becomes progressively harder to boil and remove water from (just as a similar concentration effect makes ice cream progressively harder to freeze).

If we get to 85% sugar or so, we can make a soft toffee, but going closer to complete water removal will give harder products like boiled sweets, candy, and butterscotch. Cooling the product then increases the viscosity and "fixes" the texture, but the product is so water depleted at this point that it can even absorb moisture from the atmosphere if conditions are not kept dry and soften on cooling as the sugar becomes even slightly diluted.

With many components of food, when we talk about turning into a solid, we mean crystals. When we make ice cream, we turn water into crystals (ice), because the water has been brought below the point at which it melts (on heating) or freezes (on cooling). Butter is solid when taken out of the fridge because a certain proportion of the milk fat is below its melting point and is in solid crystals, like water being converted from a liquid state to a solid one by temperature alone.

The case of sugars is somewhat different though, as the key factor is concentration rather than temperature (although the two are linked), and the fact that, as previously mentioned, a certain amount of liquid can hold in solution only a certain amount of sugar; go beyond this limit of saturation and the additional sugar that cannot be dissolved forms crystals. The interaction with temperature is because hot water can dissolve more sugar than cold water; start with water close to boiling and we can dissolve a lot of sugar (as when we make jelly), which we could depict as the highly energetic water molecules bouncing around and keeping the sugar molecules constantly buffeted and moving too fast to make crystals. Cool the water, though, and we remove energy from the system (literally and thermodynamically) and the sugar molecules move progressively more slowly, so that collisions will result in cooperation rather than recoil. So, what was initially a hot solution rich in sugar will, when cooled say to room temperature, turn into a liquid phase (the water with exactly the amount of sugar it can dissolve at room temperature) and a solid phase, which is the rest of the sugar. The result of this game of molecular musical chairs is that when the music stops (in the cool state) lots of sugar molecules are left without a chair and leave the game.

Interestingly, like children at the end of that game who don't know what to do when they find themselves without a chair, there is a period of uncertainty as the newly insoluble sugar molecules act as if unsure of what to do or how to behave. The mixture is at this point referred to as supersaturated, and some additional input is needed to prod this into forming crystals, and sometimes to direct the form of crystals to be formed. This could be provided by something as simple as the presence of impurities such as dust particles or air bubbles in the sugar mixture, which provide points of disruption around which crystals can cluster (what is called nucleation), like a zone in the mix in which aberrant behavior seems to be allowed and where the sugars feel free to break the rules and form crystals.

Another example of how crystallization of sugars can be directed is seen when whey, the by-product of making cheese, is concentrated to very high solids levels. Whey is initially a very watery liquid (around 94% water), but most of the nonwater components are the milk sugar, lactose (around 80% of the whey solids). Arguably the most valuable component of the whey is the protein, even though there is a lot less of it there than there is lactose. In processing the whey into concentrated or dried products, perhaps after removing the protein, we need to mind the properties of the lactose very carefully, as these can determine exactly what kind of end product results.

Compared with other sugars, lactose is complicated because it can form several different types of crystal, which are due to tiny differences in the arrangement of atoms and chemical groups within the lactose molecule. In a clear demonstration of the power of chemistry, lactose can exist in two interconvertible forms, called alpha- and beta-lactose, which differ only in the exact arrangement in space of three atoms (two of hydrogen and one of oxygen) within the molecule. If these are arranged one way relative to the rest of the molecule, it is called alpha, and if they point the other it is called beta. This sounds like an astonishingly minor change but, for those who dry dairy products in which lactose is a major constituent, getting the conditions right so as to have one or the other form is massively important and can make the difference between a powder that after months of storage can flow freely and be scooped, measured, dispersed, and poured easily, and one that has turned into one massive lump of rock-like uselessness; flipping that one chemical bond's orientation in space can make all that difference.[9]

The significance of these different forms of lactose crystal means that, when crystals do form, processors need to be exceptionally careful to get the right kind. So, if whey is evaporated to remove a high level of the unwanted water present, a hot solution rich in lactose forms, but as this cools the excess lactose needs to be directed to form the desirable crystal form. In practice, this is achieved by the simple step of adding to the mixture a small level of the right kind of crystals. This is as if the lactose present is unhappy, knows it needs to form a crystal, but doesn't know which to form. When we show it a clear example, though, it goes "yes, that'll do me" and copies the

---

[9] This is referred to as caking, but somewhat ironically it is not good for making cakes!

proffered example exactly, giving the right outcome and hopefully eventually stable powders and other products. Clearly, lactose, like some people, is strongly influenced by the example of others and is happy to follow the direction of the crowd.

The ability of lactose to form crystals and solid structures is often used to good effect in a very common application of all the lactose produced as a by-product of the manufacture of dairy products such as cheese and casein. Lactose is the solid carrier material within which vanishingly small quantities of biologically active pharmaceutical molecules are carried into the required site within our bodies in medicines provided in the form of pills or capsules.

Going back to the idea of sugars as crystals, just as some sugars can form different types of crystal, so the properties of high-sugar solid products like sweets are dependent on the types of crystal formed. Looked at under a microscope, sugar crystals can take a variety of complex shapes and can often resemble beautiful jewels. In some cases, the analogy is less flattering, as when a certain type of lactose crystal is referred to as "tomahawk-shaped."[10]

Just like jewels, such as diamonds and rubies, sugar crystals are multifaceted structures that can have reflective surfaces that reflect, refract, or otherwise tease the light shining onto them to give certain appearances, from dull to sparkling. In jewels, careful manipulation of manufacturing conditions and subsequent polishing can turn a rough stone into a valuable ornament. For sugars, the polishing might be a little difficult (and weird!), but the manufacturing process has to be similarly controlled to get the best visual properties and appearance.

If the crystals that are induced to form are large, with large amounts of reflective surface, a glassy or even see-through sweet can be produced, while conditions that give much larger numbers of much smaller crystals will give a more "matte" appearance, as we might expect from a toffee or fudge. Determining which is the end result depends on factors such as the sugar level reached and the rate and program of cooling applied during solidification.

If, during manufacture, temperatures are reached that result in breaking down of the sugar molecules to yield a huge range of products with different colors and tastes, we can get caramels. This usually requires temperatures close to 200 °C, and in baking biscuits or cookies this is the temperature at which distinctive colors and odors that can be directly linked to the caramelization reactions can be detected. The threshold around this temperature is even reflected in recipes that demand such temperatures for food that is to turn visibly brown, whereas those that are to have a lighter color will be cooked at lower temperatures. Different sugars caramelize at different temperatures though, and the fructose in fruit and honey will start to undergo such reactions at lower temperatures than sucrose, and so adding honey to baked products can result in browner colors at lower temperatures.

---

[10] The Native American axe-like weapon, not the larger ballistic missile version, but the concept of damage (large or small scale) is key, as such crystals have sharp edges, like the axe, and give a grainy or sandy mouthfeel when food in which they are present is consumed. I guess this is still better than consumption leading to the sensation of a large explosion.

One of the most important reactions for developing color (browns normally, but also possibly reds and oranges under alkaline conditions) is the Maillard reaction, and the addition of milk powders to sugar mixtures helps nice dark colors (and creamy dairy flavors) to develop in such confectionaries. Whether a particular product undergoes more Maillard reactions or caramelization reactions will depend on facts such as acidity or alkalinity (as Millard reactions are favored at higher pH values) and the type and relative proportions of sugars and proteins present.

Continuing this scientific ramble around Ye Olde Sweete Shoppe, we come to the jellies and gums, like wine gums. We saw earlier that there are few molecules as deft at controlling the mobility of water and manipulating texture as polysaccharides or gums, and in such products the presence of such molecules means that the heavy lifting of structure building doesn't have to be borne by simple sugars and we don't have to physically remove the water by evaporation or boiling, as we can instead just trap or absorb it using polysaccharides, and so adding these molecules can give us softer, chewier structures, into which we can add whatever flavors and colors we can dream of.

Finally, within this discussion of mouth-watering sweet treats, we should mention chocolate; the properties of chocolate though, in contrast to the products previously described, depend much more on lipids than on sugars. There are some products, though, in which sugars and the lipids of chocolate combine, for example to give products with a chocolate exterior and a soft and flavorsome center.[11]

To produce such products can involve the exploitation of another property of sugars, mentioned before for them and several other molecules in food, in that they can be converted from one form to another, often with very different characteristics, by the action of enzymes. In these cases, a solid large mass of sugar (sucrose, table sugar) can be produced to which is added an enzyme called invertase. This mass is enrobed in molten chocolate, poured around a solid core shape (picture a sweet sphere of sugar plus an embedded enzyme around which chocolate is poured to form, on cooling, a lovely smooth shell). During storage, the invertase enzyme starts to act and breaks down the sugar molecules, progressively liquefying the core, but having no effect on the encircling chocolate wall; on consumption, hey presto, a chocolate exterior is bitten through to release a soft center within.

So, sugars, large and small, can have roles in our food far beyond what we expect, and while our sweet tooths may appreciate them most obviously, there is a lot more that they contribute, and removing or reducing their levels might have more consequences than the nutritional.

---

[11] Such as the wonder known as the After Eight in many countries, where a dark chocolate envelope bears a soft message of delicious mint within.

6

Fats, Oils, and the Uneasy Truces of Emulsions and Foams

The magical structural properties of fat in food

Of all food constituents, perhaps that which receives the worst press is fat. The merest mention of fat in food tends to be regarded as a negative thing, and for years the advice of nutritionists has tended to uniformly focus on its avoidance or reduction in our diet. While recent years have perhaps seen a reevaluation of the relative importance of fat, and its reputation may be undergoing a gradual thawing, my focus in this chapter is not on the controversies of its merits or otherwise from a nutritional perspective. As should be clear by now to anyone reading this book, neither my expertise nor my focus is in such realms, and so, for now, without fear or favor, I consider the properties of fats from an impartial perch, in their own regard as key constituents or ingredients of many food products, whether we like it or not.

The picture of fat I want to paint here is one of a material that, before it enters our bodies to exert whatever physiological effects it may, hugely affects the properties of many food products, and in some cases can have effects on food texture even more significant than those of proteins and polysaccharides described in earlier chapters, where the level of fat present is sufficient to allow it to dominate the properties of the food.

Why is this?

Imagine for a moment a hypothetical magic material that determined texture and consistency of food in a way that was dependent on temperature such that it changed its properties dramatically when exposed to different temperatures regularly encountered by food. Maybe such a material would be solid, or almost so, at refrigeration temperatures, but melt at almost exactly the temperature of our mouths when we consumed the food, to soften and make our food easier to chew, while, at higher temperatures yet, as it might encounter during processing or cooking, it became a liquid, which could flow and move and be molded and reshaped, or even divided up, as we saw fit.

This magic material might also have an interesting relationship with heat in different ways, in that it opposed the flow of heat (in some cases a good thing, in others not), while its melting actually absorbed heat to fuel its liquefaction, giving it a cooling effect when eaten.

Then, when present in food systems alongside other constituents like protein and polysaccharides, which might otherwise form solid structures that might be a little too rigid or crumbly, this material might insert itself into these structures to break them up and provide soft squishy interruptions, toning down the texture and making it more pliable, while also resulting in a dynamic structure that changed as the temperature changed.

This sounds like the kind of material that, if it didn't exist, food scientists should seek to invent. Luckily, however, we don't have to invent it, as this magic material exists; it is the fat we find in many food products, and the properties I have described are essentially exactly those of fat in milk and dairy products. Other versions of this magic material also exist. Some forms of fat, which we call oils, are liquid across the complete range of temperatures we encounter in typical food systems, even at freezing temperatures.

Whether fat or oil, liquidity is important in a whole range of food applications. For example, when we think of a machine, full of dry parts that need to mechanically move in a smooth and reliable way, often while in direct contact with other dry parts, we help them to move more easily by lubricating them with something that reduces friction and encourages smooth motion. Similarly, we have for centuries used oil of a food rather than of a petrochemical nature to lubricate dry and otherwise difficult-to-chew materials, such as starchy bread, into a more palatable form.

The chemical basis of fats and oils

To understand fat better, we need to first consider what it is, and why, for example, that which we find in vegetable or plant sources is usually a liquid (oil), while that from animal sources (fat[1]) is typically relatively solid, at least at room temperature or below.

The key molecules in fats and oils are called fatty acids.[2] These are chemically composed of long chains of carbon atoms, most of which are chemically bonded either to other carbon atoms or to hydrogen atoms. All molecules are built up from simpler atoms of elements such as carbon, hydrogen, and oxygen, such as a water molecule being built from one oxygen atom and two hydrogen atoms, and the linkages between the molecules are called bonds (of which there are different

---

[1] In this chapter, I use the generic term fats and oils where they make sense, just as we don't think of olive fat or milk oil.

[2] Because oily acids just sound weird.

kinds, but we don't need to get into that level of detail). Each atom can, because of its atomic structure, form different numbers of bonds with other atoms. Looking at water again, we can see that an oxygen atom has two bonding "spots," while a hydrogen atom has one, and this gives us the ratios of each found. Carbon can form four bonds.

In a fatty acid molecule, most carbon atoms are chemically bonded to two neighboring carbon atoms, plus one or two hydrogen atoms. This structure results from the atomic structure of the carbon atom, which can be imagined as a cross with two horizontal and two vertical arms. In a fatty acid, the horizontal links are to other carbon atoms, while the vertical ones are to hydrogen atoms. On this basis, we can picture a chain of carbon atoms linked like C–C–C–C–C, with each of the C's having two hydrogen atoms pointing upward and downward.

Fatty acids are classified on the basis of how many carbon atoms are found in the chain, and can be termed short-chain (maybe 4–6 carbons), medium-chain (10–12), or long-chain (16 or more); the chain in all cases is terminated after the last carbon in the chain by what is called a carboxylic acid group (which gives the "acid" part of the term fatty acid).

The chemical nature of this repeated carbon–hydrogen–carbon structure is very uncharged in electrical terms, with strong, stable bonds. This kind of structure doesn't form bonds with or attract water molecules, and so water is effectively repelled from the chain. This is why, as the old saying goes, "oil and water don't mix because of the inherent hydrophobicity of the hydrocarbon chain" (sometimes this saying is truncated a little).

To make things a little (more) complicated, though, the type of bonds between the carbon atoms can differ. In the structure I have just described, the carbon atoms are linked to each other by what is called a single bond (only using up one out of the four bonding "spots" a carbon atom has to bond to another carbon atom). Sometimes, however, the carbon atoms form an even stronger bond and double up their commitment to each other, forming a double bond that, for each atom thus entangled, uses up two of their bonding "spots." In such cases, there is one fewer bonding spot left over to share with hydrogen atoms, and the carbon atoms in these double bonds are linked to one hydrogen atom each only.

In such bonds, two conformations or arrangements of the hydrogen atoms around the double bond can be found, called *cis* (where two hydrogen atoms are pointing in the same direction on one "side" of the bond) and *trans* (where they are on different sides and pointing in different directions). At a molecular level, the fatty acid molecules have slightly different shapes if the bonds are in the *cis* or *trans* form, but critically *trans* fats result from processing of fats and oils (called hydrogenation) and are less commonly found in nature. There is ongoing concern as to whether *trans* fats could be harmful to human health.

Fatty acids that have only single bonds between carbon atoms are called saturated fatty acids, because the bonding spots for hydrogen are saturated and no more hydrogen can be added. Fatty acids with double bonds, on this logic, are called unsaturated fatty acids; if a fatty acid has just one double bond along its chain of carbon atoms it is termed monounsaturated, while if it has more than one it is called polyunsaturated. This chemical structure difference has profound implications for the properties of fats and oils and is a great illustration of how minor molecular changes can have massive real-world implications for the character of food. In general, animal fats are rich in saturated fat, while plant oils are rich in unsaturated fats.

It might sound like there a huge range of possible fatty acids, based on numbers of carbon atoms, numbers of double bonds, and also where these double bonds are placed in the molecule (between which atoms). In practice, however, only a reasonably limited pool of fatty acids is found in foods. In milk fat, for example, there are maybe a dozen major fatty acids, some of which are long-chain and saturated (such as palmitic acid and stearic acid), while others are short-chain (butyric acid) and others are long-chain and unsaturated (such as oleic acid). The names of some of these fatty acids reflect where they are commonly found, with olive oil being rich in oleic acid and palm oil being rich in palmitic acid.

One of the most important properties of a fatty acid for its properties in food is the temperature at which it turns from a solid to a liquid (its melting point). All fatty acids will solidify at some key temperature, and form crystals below that point, while existing as a liquid above that point. A key difference between saturated and unsaturated fats is that the former tend to have high melting points, while the latter have low melting points. This is why vegetable oils are typically liquid, while animal fats are generally at least partially solid at the temperatures at which we use them in the kitchen, like in the fridge or on the table.

Now, any fat or oil is a mix of different fatty acids (and the triglycerides in which they are found, as will be discussed shortly), all of different melting points. So, when we heat or cool an oil or fat the transition from solid to liquid can be quite gradual, as different fatty acids progressively cross the threshold at which they change their state. This can be much more gradual compared with the abrupt changes we see when melting or freezing water, for example, where all the molecules change their state at the same temperature.

In the case of milk fat, which contains a quite high level of saturated fat, at refrigeration temperatures (typically 5 degrees Celsius [°C]) the fat is a mixture of solid and liquid (all fatty acids with melting points below 5 °C are liquid, while those with melting points above 5 °C are solid), but the solid character dominates, and the butter is hard to spread straight from the fridge. If it is taken out of the fridge and allowed to stand at room temperature (say 20 °C), all the fatty acids with melting points between 5 and 20 °C liquefy, the proportion of solid fat decreases, and the butter softens notably. Put it on toast then, and it warms up a little more and some more of the fat liquefies and so forth. To melt all the milk fat, we need to heat it to temperatures above around 40 °C.

In the case of oils, however, unsaturated fats dominate, and these have much lower melting points than saturated fatty acids (often well below freezing). So, at room temperature, or even at fridge temperatures, the fatty acids are all on the liquid side of their melting point, and the product remains free-flowing and unsolid, and we buy vegetable oil in a bottle, rather than in a solid block like butter.

This difference explains very simply the differences between traditional butter and the products known as spreads. To get over the main disadvantages of butter (that it is hard to spread from the fridge and contains saturated fat, which is generally believed to be less healthy than its unsaturated counterpart, and is also more expensive than vegetable oils), products were developed in which some or all of the milk fat was replaced by vegetable oil; this gave a cheaper, more spreadable, and healthier product in one fell swoop!

To think about the structure of fat-containing products a little more, it is important to consider what we mean by solid fat in a product. Analogously to water, when fats solidify they form crystals, which is what causes the physical change from liquid to solid. So, in a product like butter, at a given temperature some of the fat is solid (in crystals) and some is liquid, and this is what determines how a knife will go through it and how it will spread on bread or whatever. In preparing the butter, however, it is possible (by careful manipulation of cooling rates and temperatures) to change the nature of the crystals that form, and hence the structure of the final product. Put simply, even if two samples of butter contain 50% of their fat as crystals, one in which those crystals have been induced to form a large number of very small crystals will be softer and more spreadable than one which contains a smaller number of larger crystals. When we then subject butter to physical force when spreading it on a bread surface using a knife, one simple consequence is that we push the crystals apart, so that they no longer interact and the product becomes softer.

Besides spreadability and textural differences, one of the most significant differences between saturated and unsaturated fats is their resistance to a chemical process called oxidation, which involves attack by oxygen that might be present in the atmosphere surrounding the fat-containing product. Oxidation can result in the development of off-flavors in such products, and rancidity in particular, and so is clearly undesirable, and products containing high levels of unsaturated fats are prone to such problems. In a related example, fish tend to contain quite a lot of unsaturated fat, which helps to keep their bodies warm in water without solidifying, but makes fish flesh susceptible to oxidation, which can give us some pretty nasty smells.

Luckily, there are reasonably straightforward steps that can be taken to reduce the risk of such problems. These include the inclusion of ingredients that oppose oxidation reactions (antioxidants[3]), and the avoidance of factors that promote oxidation, such as the presence of certain metals, access to light, and obviously the presence or access of oxygen. Consideration of such factors makes it very easy to

---

[3] Key examples of such ingredients are certain vitamins, which coincidentally are often added to unsaturated-fat-rich spreads (considered alongside their nutritional benefit, this makes them dual-purpose food ingredients).

see why products such as butter and spreads are packaged as they are (in light- and gas-impermeable packaging, for example).

From fatty acids to triglycerides: Safety in numbers

While fatty acids are the major constituents of fats and oils in food, we rarely (but not never) find them in a free form. They are most commonly found bonded to another molecule, an alcohol called glycerol. An alcohol is defined chemically as a molecule that contains within its structure a chemical group consisting of an oxygen atom and a hydrogen atom bonded together (an OH or hydroxyl group).[4] This chemical group is extremely water-loving (hydrophilic), which is why water and alcohol mix readily, but not necessarily why alcohol can get people mixed up. The same OH group is a defining part of alcohol molecules, such as the ethanol found in all alcoholic beverages, from the very cheap (OH crap) to the very expensive (OH dear[5]), and is perhaps responsible for the feeling encountered the morning after excessive consumption of hydroxyl groups (OH my head).

In chemical terms, acids and alcohols bond together to form a molecule called an ester.[6] Fatty acids and glycerol accordingly form what are called ester linkages. Glycerol actually has three of these OH groups, and so is like a molecular docking station onto which up to three fatty acids can bond. If one fatty acid is bonded, the resulting molecule is a monoglyceride, if two a diglyceride, and, if three are bonded, it is a triglyceride. The more fatty acids are docked, the more the naturally exuberantly water-worshipping nature of glycerol is dulled down, and triglycerides have essentially completely lost their affinity for water, becoming completely hydrophobic.

The glycerol, by "consuming" fatty acids, has become fat, and the consumption of the resulting compounds in large doses passes on the effect, like molecular contagion.

Monoglycerides or diglycerides, however, retain a mixed character, with some hydrophilic behavior, which is due to their remaining one or two free hydroxyl groups, and some hydrophobic behavior, which is due to the fatty acid(s) stuck on. Molecules that have this type of schizophrenic character are called amphiphilic and, as we will see in later chapters, are actually very useful specifically because of their ability to mediate between oil and water, which is why many food package labels contain the words "emulsifiers (mono- and di-glycerides)."

So, most fats and oils are largely composed of triglycerides, and these are actually typically a very mixed bunch of molecules. While it was stated earlier that milk fat contains around a dozen main fatty acids, these can combine in triglycerides in

---

[4] Try asking for something with a lot of hydroxyl groups in it next time you are in a bar and please let me know how you get on.

[5] Not to be confused with expensive mineral water, which is *eau* dear.

[6] I have always wondered if a lot of people with a similar-sounding name were born in the 1960s as the product of a reaction between acid and alcohol.

a huge range of options. For example, if there were only three fatty acids (A, B, and C), glycerol molecules could bond with them in combinations such as AAA, AAB, ABA, BAA, ABC, CCC, CBA, and so forth; it should also be noted that the order in which they line up on the glycerol molecule is important, and so AAB and BAA are not interchangeable.

So, while a food product may contain a relatively limited number of fatty acids, it contains a massive variety of triglycerides, and the ultimate melting behavior of the fat depends on the overall profile of triglycerides, as well as on individual fatty acids.

Fat also links to other ingredients to produce new kinds of structures and textures. We learn as a cliché that oil and water don't mix, but many foods are in part at least an emulsion, where these supposed enemies coexist happily, thanks to the presence of some mediating ingredient (sometimes our good friends the proteins) to keep the peace. Oil and water are like hostile countries who want to go to war, and only the United Nations–like peacekeeping of these ingredients keeps this from happening.

Lots of food products are actually stable systems in which oil and water have found a way to coexist and remain mixed without separating; examples include cream, milk, butter, and mayonnaise. What these all have in common is the presence of a third ingredient or component, which is either naturally present or has been added, with amphiphilic properties.

If we look at milk under a light microscope, we will easily see large droplets floating around, which are the droplets of fat. However, a much more powerful microscope would reveal how they remain floating around in the liquid without coalescing and separating into a layer; this is because of a very thin layer of molecules coats the surface of each oil droplet. This layer contains the amphiphilic emulsifier molecules, each of which has found a happy home with one foot (molecularly speaking) in the oil phase and the other dangling out into the water phase. There are a lot of other very complex molecules, including proteins and enzymes, present in this so-called milk fat globule membrane, but for now we just need to know that the reason that oil and water somehow have been allowed to mix is the presence of these emulsifier molecules.

So we have two systems of very different properties, which essentially hate each other, and need, to achieve any kind of stability, to be kept separate by a protective barrier.

Making emulsions in food

Let's consider that mixture of oil and water that we can't whisk into a stable system. All we need is to add something that contains amphiphilic molecules, and the whole picture changes. One example is egg, which contains molecules called phospholipids[7] that will do just the job; add some egg, repeat the mixing, and with enough effort we find

---

[7] A lipid (oily) molecule with a highly charged group including phosphorus attached; the lipid bit

that we now can disperse the oil successfully. What has happened at a microscopic level is that the oil, as before, was dispersed by the physical action of the whisking breaking it up into unstable droplets that wanted to come together to separate. If enough emulsifier is present to coat these droplets really quickly and effectively, though, they will get coated before they got a chance to coalesce, and so remain separate.

Not all emulsions are equally successful in remaining stable though, and the really strong drive of oil to separate from water, usually by being upwardly mobile, needs very good emulsifying power to overcome. A principle called Stoke's Law describes how quickly the droplets in an emulsion will separate, and it shows how different properties of the droplets, or the liquid medium in which they float, will affect the rate of rise or fall of the droplets. According to Stoke's Law, the bigger the droplets, the less viscous the surrounding fluid, and the bigger the difference in density between oil and surrounding fluid, the more quickly the droplets will rise.[8] The equation by which Stoke's Law is expressed also includes a term for gravity, which can be replaced by a much higher factor if we subject the emulsion to extremely rapid rotation, in what is called centrifugation.

To explain the emulsions in food a bit more, think about milk (which I do a lot), which is rare example of a food emulsion found in nature, as opposed to being created. Milk is intended to provide baby mammals with a lot of nutrients, some of which are lipids (fats and oils) or are happiest dissolved in fat (fat-soluble vitamins). To keep this stable enough to get from mother to baby/cow to calf, milk must thus be an emulsion, but it doesn't need to be a very stable one, as the time from production to consumption is very quick, so unprocessed raw milk is basically a pretty sloppy emulsion with big droplets.

If, on the other hand,[9] we snatch cows' milk away from its intended consumer and leave it stand outside the cow, we see that the globules quite quickly rise to the top and give a layer not of separated oil (the barriers around the individual globules of fat keep this from happening) but a layer that is enriched in these big droplets, while the remainder of the milk is depleted of fat; this gives us cream and skim milk. This even happens to an extent in the cow's udder, and the milk during a milking operation gradually becomes richer in fat, as the fat globules that have risen under gravity as the cow stands or walks vertically are gradually drawn downward and out.[10]

---

gives the hydrophobic part of the structure, while the charged group attracts water and makes that bit of the molecule hydrophilic. Such a molecule really has two different characteristics in one, and could be visualized as a round head with a long squiggly tail sticking out of it. The tail is the hydrophobic bit, and this wants to find an oily material to associate with, while the round head is like a beacon for the water with which it feels the greatest affinity.

[8] The law holds for any material dispersed in another one. On the one hand, if the dispersed material is an oil droplet that is lighter (less dense) than the surrounding water, it will predict the rate at which it will rise; if a stone is immersed in less-dense water, or even air, on the other hand, it will predict exactly how quickly it will sink.

[9] Or, in the traditional method, using two hands and a bucket, and trying not to spill it on either hand.

[10] This could be avoided only if we somehow agitated the cow before milking to mix its milk, but of course then we would get a milkshake.

So, milk readily separates under gravity, and if we speed up gravity by centrifuging the milk, taking advantage of Stoke's Law, we can in seconds prepare streams of cream and skimmed milk, which we can make into different dairy products or remix to give us products of defined fat content somewhere between the two separated extremes.

To briefly introduce here the principle of centrifugation, if you picture taking a bucket of milk and attaching a rope to it, and then swinging it around your head very rapidly, the milk would not fly out because the circular motion will fling it outward, toward the bottom of the bucket. The skim milk part is heavier and will be thrown out farther than the lighter cream. So, when your arm got tired and you looked into the bucket at rest, you should see the cream nicely separated to the top of the bucket. In practice, in laboratories and milk processing plants (and even a few kitchens), this is made less potentially messy by using a machine called a centrifuge or separator that mechanically spins tubes or containers of emulsions or mixtures of solids and liquids to speed up their separation.

Because one of the main drivers for separation is the size of the fat globules, we can apply mechanical force to improve the emulsion by physically breaking the fat into smaller globules, which then separate much more slowly. This is the principle of homogenization, whereby the milk is forced under pressure through narrow gaps that basically batter the droplets into mechanical rupture, shattering each into smaller droplets. An analogy would be forcing a basketball through a narrow hole and ending up with tens of thousands of table tennis balls out the far side.

Of course, reducing the size of the droplets hugely increases the surface area that needs to be stabilized, and there needs to be enough emulsifier amphiphilic molecules present to coat all of this new surface. If there wasn't, the droplets would simply come back together to their original size. In milk, the extra surface gets coated in proteins that fulfill this function.[11] This also makes the globules denser and heavier, which further discourages them from separating.

This is what gives us homogenized milk in the cartons in which we most readily purchase milk today, and is why, when we pour out their contents, we don't see a visible cream component. It is not that long, though, since unhomogenized milk was the preference, and I clearly remember glass bottles of milk being delivered to our Dublin doorstep in the 1970s. For these bottles, a foil cap was all that, sometimes unsuccessfully, kept the rich cream layer on top protected from the attentions of magpies and other birds. A further consequence was that sometimes, on very cold days, that top layer would be quite solid and needed to be punctured to allow the milk to flow.

This observation raises a really important point, which is that the physical state of the fat in food is critical when processing it to change its properties. For example, to separate or homogenize milk, we need the milk fat to be liquid. At fridge

---

[11] As introduced in Chapter 3, the caseins and whey proteins have enough amphiphilic (mixture of fat- and water-friendly) character to allow them to act at the interface and stabilize the emulsion.

temperatures, as previously noted, much of milk fat, because of its saturated nature, is actually in a solid crystal form (think of how hard butter is straight from the fridge). When we warm butter, some of the fat gradually melts, as individual components of the fat reach their individual melting points, and at around 40 °C all the milk fat is liquid. This means the droplets will flow and stretch and rupture when we want them to, far more readily than if they were solid; to illustrate this point, think about which would be easier to flow or divide or spray—ice or water! Thus, to treat milk when the goal is to change the properties of the fat, such as separating or homogenizing it, we need to have the milk quite warm or the processes just won't work very well. If we were using oils that are liquid at much lower temperatures, like a mayonnaise containing vegetable oils, we can work at room temperature and still get good results.

In the case of mayonnaise, a further factor influencing the stability of the emulsion is the presence of vinegar, which makes the environment more acidic, under which conditions the droplets of oil (or more precisely the egg proteins on their surface) become more charged, and hence mutually repulsive, which helps to keep them apart and from coalescing.

A microscopic image of an emulsion can be seen in Figure 6.1, showing how the oil droplets form an almost perfect sphere (seen in cross-section), but are clearly surrounded by a layer of red-stained protein, keeping the droplets stable in the surrounding liquid. This was taken using a different type of microscopy than the ones we have seen already, called confocal laser microscopy. This is probably closest in principle to a light microscope, but uses highly precise beams of laser light to study one part of a sample at a time, building up almost three-dimensional images (and cross-sections, such as this one, proving very powerful here when combined with chemical stains that make the fat shine as green, while the protein appears red).

This could easily be an image of a product like the mayonnaise shown in the left-hand side of Figure 6.2, which is nice and stable and so remains even and free-flowing. If the emulsion is disrupted, though, for example if the conditions interfere with the stabilizing proteins or the mixture was not mixed sufficiently to disperse the fat, you would not need a microscope to detect the difference, as the right-hand side of Figure 6.2 shows.

Figure 6.3 shows more confocal micrographs of a different kind of dairy emulsion, in this case cream cheese. In this case, the protein network is again stained red and is composed of coagulated casein, the protein from milk, which has been transformed into the semisolid structure of the cheese by the processes discussed in Chapter 4. Dispersed within this protein network, though, are green droplets that are the fat globules, which form a stable emulsion thanks to homogenization of the milk before the cheese was made. Comparing the top and bottom pictures in this image, though, shows some significant differences in the same cheese made with different fat contents, with far fewer green drops in the bottom and much larger spaces and voids within the protein structure (the dark chasms). This is a lovely example of how much fat contributes to the structure of food, and how its removal or reduction really alters the fundamental properties of how the product is built, and often gives textures that (besides obviously being less creamy) appear weaker and more "watery" (as the ability of protein to control strength is reduced and the properties become more dominated by the water present).

From cream to butter

As we have just seen, the droplets of fat in milk and cream are not too tough and can be easily damaged by mechanical force, as we do when we homogenize milk. Indeed, the production of another ancient dairy product is based on this very principle, and we have known this art for many millennia before the science was understood.

Butter is a product with a long and proud tradition in many countries, including Ireland. Not infrequently, wooden containers of primitive butter are excavated in Ireland's large expanses of cool peat bogs, where they have been buried as far back as tens of thousands of years ago. Similar practice has been observed in Scotland and elsewhere, but the reasons remain mysterious (preservation? storage? a strange game of hide-and-seek for children's parties?). Samples from museums have even been analyzed using highly sensitive modern analytical techniques, as reported by Irish scientists in the *International Dairy Journal* in 2007, and found to contain familiar molecules to those we would find today, simply in more degraded form (the scientists did not appear to have gone as far as to taste the samples, presumably only because there wasn't enough available).

In more recent centuries, farmhouses in Ireland commonly possessed a small wooden churn, often shaped like a cylinder lying on its side, into which the cream that naturally separated when raw milk was allowed to stand was poured, allowing plenty of space for air (the important invisible ingredient). Plenty of simple brute force was then applied, usually by rotating the churn vigorously using a handle, or perhaps by attacking the cream viscously using a long wooden paddle or dasher.

A few minutes of this butter-battering was usually sufficient to whip the cream into a thick viscous mass, which was significantly expanded in volume from the original cream, followed (presumably after a short rest for the butter-beater) by continued exertion. This would lead eventually to the thump-thump sound of pieces of solid material floating around the churn, gradually separating from a surrounding liquid. These lumps of proto-butter could then be recovered by straining them out of the liquid (buttermilk), and further mixing, kneading, and squeezing them (removing more buttermilk), perhaps with addition of a little salt, to give butter.

This has been practiced for thousands of years, probably originally as a means to preserve cream in a more stable form that made the consumption of otherwise dry dietary staples such as rough bread or potatoes more pleasant, while exactly the same methods can allow you to make butter today (maybe speeding up the mechanical part by using an electric mixer, whisk, or unit like a Kenwood Chef).

This art, like so many others in the world of food, is the cultural application of what is actually a highly scientifically complex process that, technically speaking, is called phase inversion of the emulsion of cream. Again, art comes long before science catches up.

So, what happens in the churn?

The key to understanding the process is the stability of the milk fat globules and the destabilizing twin influences of physical force and air.

As discussed earlier, the fat within the globules in milk is kept separate from its archenemy, the watery phase that constitutes the vast bulk of the volume of milk, by a thin protective membrane. This membrane is fragile, however, and can readily be punctured and fragmented by application of brute physical force. This we do by the churning actions described, which simply knock chunks of the membrane off the droplets by impacts with the walls of the churn or the beater within, exposing the hydrophobic fat within. At each of these places where the membrane has been stripped away, oil and water now meet, look each other in the eye, and immediately seek to avoid each other by any means possible.

While all of this tense brinksmanship is going on, the globules are flying around at high speed thanks to our butter-maker's vigorous efforts and will frequently collide with each other. When two globules with intact membranes meet, the membranes keep them separate and they fly apart again but, when two with damaged surfaces collide, the fat has a chance to minimize its contact with the dreaded water by instead aligning with more familiar fat, and so two globules might stick together at their damaged places. As time progresses, three, four, and more join this tentative huddle, forming small chains and groups.[12]

This could go on for a while, but the presence of another agent speeds up the process significantly: bubbles of air whipped into the cream by the churning action. Air also doesn't like water, and each bubble is essentially a large hydrophobic void, the surface of which is a much more hospitable environment for a poor, wounded fat globule than the watery hell through which it is flying at insane speeds. So, the damaged globules cluster readily at the surface of the air bubbles, which eventually become coated in ever-more-complex layers of damaged fat, and neighboring bubbles can even become linked by chains of other globules. If the temperature is low, the saturated milk fat is pretty solid, and this air-entrapping network is also pretty firm, so the cream expands in volume (thanks to the entrapped air) and stiffens, which is exactly what we get when we whip cream for coffee or desserts.

In making butter, though, we go much, much further, and keep beating the whipped cream. The lumps of fat surrounding the bubbles grow and grow, and eventually the bubbles literally burst from the pressure, and we get lumps of

---

[12] Like hydrophobic humans seeking shelter by huddling together under a large umbrella or tree as the rain gets heavier and heavier.

extracted fat, called butter grains, which were the output of the churning described earlier.

As well as auditory clues to this process happening (as the growing butter grains knock off the walls of the churn), this process can be even visualized through color as well. Cream is white, due to the ability of the fat globules to obstruct the path of light through it, and whipped cream is likewise white. However, milk fat is yellow, to extents that differ based on the diet of the cow that produced the original raw milk. The yellow color is due to a molecule called beta-carotene (yes, it is also found in even more concentrated form in carrots, hence the name), which comes originally from grass. The more grass the cow eats, the more beta-carotene is present and the more yellow the milk fat. In Ireland, cows eat lots of grass because our historical farming habits, which favor milk production over the grass-rich summer months, and so we produce particularly rich golden milk fat. The reason then that the milk doesn't look yellow is that this natural pigment is hidden behind the milk fat globule membrane and cannot be seen. Break this membrane, however, and you extract the fat and in parallel the yellowness, and the progress of effective butter-making toward the desired conclusion is clearly evident from the gradual appearance of this proud regal color.

The buttermilk is then drained off, and further kneading and mixing break down the borders between the grains, and work the butter into a homogeneous mass, just as individual balls of dough can be kneaded into a single whole while baking. This isn't pure fat, though, and a small percentage (around 15%) remains of the watery buttermilk dispersed through the butter. Peer at butter down a microscope and you can see tiny round pools of this water, dispersed as droplets throughout the continuous fat phase.

So, to summarize this bit, we started with cream, which is an emulsion of oil (fat globules) dispersed in water (skim milk); this was then turned in into butter, which too is an emulsion, but this time of water (buttermilk) dispersed in oil (the butter fat). One emulsion, turned inside out by the long-known application of elbow grease, through another highly scientific phenomenon understood culturally long before the science was elucidated: phase inversion.

From real cheese to processed cheese

Other types of dairy products besides cream and butter are emulsions. For example, a long-established relation of traditional cheese is processed cheese, the kind we find in spreadable form in tubs or, in more solid form, in slices, blocks, triangles, sausage shapes, and even occasionally cans.

This product was initially a way to use up cheese that did not meet exacting specifications for composition and perhaps flavor, by melting it down and rebuilding it as a new structure in a product that allowed greater forgiveness for slight deviations from idealness than allowed for in conventional cheese.

If we compare the textures of natural and processed cheese, we tend to most commonly apply the adjective "plastic" to describe the latter, a word we should hopefully never apply to a fine block of mature Cheddar. In one case, we envisage a smooth homogeneous structure, in the other a more friable and brittle structure (the former being termed long, the latter short, the length reference relating to how far you could bend or stretch the material before it broke).

Processed Cheddar cheese is made from natural Cheddar, but the process of transformation of one to the other requires a reengineering of cheese structure. To understand what happens, picture what would happen if we melt natural cheese to a liquid and pour it into a tall cylinder. Leave it long enough and we would likely get a solid lump of insoluble protein at the bottom, a layer of separated fat or oil at the surface, and a murky, milky liquid in between. Shake this as much as you like, and nothing resembling cheese, natural or processed, will result.

Now, what kind of a stable food-like system could we make from a mix of water and oil? Obviously, this has the potential to be an emulsion, but to make an emulsion we need an emulsifier to sit at the oil–water interface and keep the peace. What present in the mixture could act as an emulsifier? The most likely suspect here is the protein, as protein molecules, because of their structures and the range of amino acids they contain, can arrange themselves at interfaces (like egg proteins do in foams, for example) and, being bulky molecules, can do a good job at stopping neighboring oil droplets from coalescing just by virtue of their physical shape and protruding parts, like a dense, thorny bush keeping the borders of a nice field free from interlopers.

However, when we convert casein in milk into cheese, the form of casein that results is a destabilised calcium-bonding form, which isn't very good at being an emulsifier, as it tends to form a networked protein structure that doesn't come apart to coat droplet surfaces very well. The calcium is the problem here, so what we need to do is to add agents that react with the calcium and peel it away from its sticky role between casein molecules. Lots of inorganic molecules (salts) can do this, particularly complex molecules containing lots of phosphorous molecules, called phosphates, and salts of fruity citric acid, called citrates.

These phosphates and citrates, in forming stable molecules, need to bind to a metal ion, and so often are found in complexes with sodium. Add sodium citrate to the calcium form of casein, though, and a sort of molecular switcheroo ensues, as the citrate decides it prefers to bond to calcium over sodium and ditches the latter to form a new compound with the former; the newly ditched sodium, on the molecular rebound, hitches up with the casein. The new protein complex formed, called sodium caseinate, is much less self-absorbed than the calcium form was and readily forms a suspension of reactive mobile molecules.

In making processed cheese, all this happens in a hot system (to melt the fat), to which the water and sodium citrate (or phosphate) has been added. An important requirement during this step is physical force, in the form of mixing or chopping blades, that disperse the melted fat into liquid droplets, which then swirl around at a high speed and bump into the mobile sodium caseinate molecules. In such a collision, the caseinate sees an environment that is preferable for it (because part of its molecular structure hates the water in which it is now dispersed) and so will stick to the globule, such that the hydrophobic parts of the caseinate are associated with the fat and the hydrophilic bits stick out into the surrounding water phase. Hereby is born an emulsion!

Leave the hot mix stirring long enough for the oil to all be emulsified in this manner, and then when it cools the mass will set into a homogeneous solid emulsion, which is our processed cheese. The combination of (relatively) low-moisture content, the pH at which the proteins find themselves, and the presence of a fairly high level of fat (which is solid at refrigeration temperatures) means that the processed cheese is solid. Of course, in a solid, the emulsion doesn't need to be as perfectly formed as it needs to be in, for example, mayonnaise; even if an oil droplet wanted to separate under gravity or coalesce with neighboring droplets, it is entrapped in a solid-like matrix that frustrates its upwardly mobile instincts.

Another advantage of processed cheese is that the emulsification process has taken place at temperatures of at least 70 °C, which has pasteurized the cheese and mixed and inactivated the microorganisms and enzymes present in the cheese. As a result, the processed cheese has a really long shelf life and is, compared with its parent cheese, almost completely biologically inert. In addition, the fact that it transitions from a liquid when molten to a solid when cooled means that solidification takes place under a highly flexible range of conditions, in terms of the mold within which the cheese turns solid, and hence the shape it adopts, whether block, slice, sausage shape, or triangle. By having higher moisture levels, we can also produce spreadable products, but in such cases the exactness of the emulsion formed needs to be higher, as mobility of the fat during storage is obviously greater.

As one key point, though, it was previously stated that cheese in this case is the source of an emulsifier, thanks to the structure of casein and its ability to form bulky layers at fat globule surfaces (the salts we add are called emulsifying salts, but all they do is convert the protein to a state where it can better fulfill this function). During the ripening of cheese, however, we know that casein is broken down into progressively smaller fragments. How does this affect its ability to act as emulsifier? Pretty strongly!

Young cheese contains lots of intact protein and hence is a good starting material for an emulsion. Old cheese contains less intact protein and hence would make a much worse emulsion, as the fragments of casein are less effective in stabilizing emulsions (or eventually, when broken down into amino acids and flavor compounds, not effective at all). However, there is a corollary to this. Young cheese hasn't been through the process by which flavor develops, and so doesn't contribute the presumably desirable characteristic of cheesiness to the processed cheese; old cheese will do this much better. Also, as the cheese ages, calcium tightly bound to the casein becomes increasingly soluble, and the proteins, being less closely stuck together, can flow more readily on application of heat, so older cheese melts more readily.

So, which do we want, structure (young cheese) or flavor (old cheese)? The answer is both, so most processed cheese products include a mix of the two, in proportions depending on the intended application of the final product. If the cheese is intended to go into a cooked product, like a burger or toasted sandwich, flavor is less important, and melting neatly is more important, so young cheese might be more represented than its more mature relative.

These principles also apply in the dish called fondue, in which cheese and wine are melted together to a delicious gooey mass into which other things (like bread) can be dipped. In this case, the wine is the secret ingredient, as it contains acids such as citric and malic acid that bind the calcium, freeing up the casein in the molten cheese to give a smooth emulsion rather than a separated mess.

Air, gas, and foams in food

We are familiar with foams in many food products, particularly whipped dairy products such as dessert cream and ice cream, as well as mousses. It is perhaps ironic that so many desserts, which might be seen as indulgent and unhealthy, actually contain so much of arguably the healthiest ingredient of all, which is air.

Air is a very popular food ingredient with food companies, because it is free and adds volume, with ice cream, for example, being surprisingly full of the stuff. Chefs love it too, and increasingly it is found beyond its traditional home on the dessert menu to appear in a starring role in savory or other dish elements or accompaniments in the form of foams. Today, in many kitchens, foams are thus being found in applications other than the sweet. Molecular gastronomy approaches, for example, have brought savory foams onto the menu of many fine restaurants, while obviously bubbles of gas are a critical element of many beverages, from the cheapest fizzy drink, through the foamy heads on a pint of beer, to more expensive champagne.

In this section, as before, my interest is in the science of how foams form.[13]

Just like oil and water, air and water don't mix, and if we bubble air into water the bubbles quickly dissipate. However, if we add the right ingredients that can grab these bubbles and trap them before they can escape, we can make a foam. The principles and steps involved in making a foam are very analogous to those previously described for making an emulsion, as in both cases we are making a product with two different phases forced to peacefully coexist, whether oil and water or gas and water.

In the first step, air (or another gas) is "added" somehow into a liquid (for example, blowing, whisking, shaking, injecting, or using a coffee milk foamer), while the liquid is being vigorously agitated so that the gas being added breaks up rapidly

---

[13] How foams form . . . a fun phrase to try and repeat multiple times really fast if you need a break from the science in this book.

into the shape in which it minimizes its contact with the surrounding liquid, which is lots and lots of small spherical bubbles.

These bubbles tend to be very short-lived, as they seek to minimize their contact with the liquid by merging together into larger and larger ones, which, being light, will strike upward for the surface of the liquid (bigger ones rising faster than smaller ones), on arrival at which they will expire in a sad little "pop." The enemy here is gravity, which can be pictured as drawing the heavier part of the system (the liquid) downward, while the lighter part (the gas) goes in the opposite direction. Water drains down, while the bubbles reach for the sky.

To retain the gas bubbles within the liquid, the key is the second step, by which something coats the surfaces of the bubbles and forms a physical barrier to their coalescence, interrupting the first stage of their escape and slowing down their upward exit. If suitably coated, when two air bubbles approach each other, the barrier forces them apart before they can merge and they pass like (gaseous) ships in the night.

The properties of the liquid surrounding the bubbles also play a key role, as we don't want this fluid to drain too quickly from the spaces between the bubbles, which would bring the bubbles closer together and make them more likely to merge. If this liquid is sufficiently viscous or entangled by the presence of other large structures (maybe fat globules or complexes of proteins or polysaccharides), this will slow down the bubbles' movement, and they will remain trapped in a stable foam.

To consider how important the liquid is, consider the fact that pure water cannot hold bubbles. We can use this fact to determine the purity of a sample of water by testing its ability to hold bubbles; if the water can froth, there is something present that is allowing it to do so. To make sparkling water, we inject carbon dioxide into water, but then we need to trap it there by pressure to stop it escaping in the absence of something to entrap the bubbles. We will discuss phenomena such as evaporation in chapters 9 and 12 in more detail, but for now let's just picture pressure as the force pushing down on the surface of the liquid. For the bubbles to escape, they need to burst at this surface, to allow their contents to escape into the vapor above. If there is a significant force pushing down on the surface, this will oppose the escape, and the bubbles will remain trapped in the liquid, with nowhere to go, and so we have essentially forced the carbon dioxide to remain dissolved in the water. Reduce the pressure, however, by opening the bottle and removing the force, and we have created conditions far more amenable to evaporation while reducing the solubility of the carbon dioxide in the water. The product then fizzes as all that carbon dioxide turns into bubbles, which then by virtue of their being lighter than the surrounding liquid will head for the surface and escape at last (hopefully slowed down enough to give us an effervescent hit when we drink the beverage).

It is a fact that all foams are unstable and will eventually collapse, as seen readily from close observation of the head on a pint of beer sitting on a table.[14] The key

---

[14] Strangely, this experiment rarely goes on for long enough to observe the foam collapse fully though!

question is not whether a noncollapsing form can be created, but rather how much that collapse can be slowed down. Bubbles are always ephemeral, doomed to a short but visually impressive life ending in a sad pop, and the life span of a bubble before this happens depends on factors such as the size of the bubble, what is present at its surface, and the composition of the fluid that surrounds the bubble.

Molecules likely to be good at stabilizing a foam will share many characteristics with those that are good at stabilizing an emulsion. In other words, they will be amphiphilic, with a mixture of water-repelling hydrophobic character (for which any material that isn't water is a friendly home, and this can be air as well as fat or oil) and hydrophilic character, which is happy in the non-air watery part of the foam.

Picture sitting on a warm day by a pool's edge, perhaps engaging in an experiment in which you monitor the rate of collapse of the foam on a glass of very cool beer, with your legs in the water, kicking idly. You in this case are like an amphiphilic foam stabilizer, with part of you in air and part in water, happily in balance in between.

Proteins are among the best molecules for this purpose, as we have seen when we discussed their role as emulsifiers previously. So, protein-stabilized foams are common in food and cooking, particularly where we use egg as the source of protein, and then coagulate the egg into a solid barrier around the air bubbles, as in a meringue or soufflé.

When we whip eggs, we entrap air, and cooking then coagulates the egg (by denaturing the proteins therein) while entrapping the air within the now-rigid protein structure. The peculiar shape of whisks used for whipping (whether manually or in automated form) is actually derived from careful consideration of the shape that will be most efficient in pulling air into the mix and creating bubbles, which are then covered by the proteins that have been unfolded into a more "sticky" state by the physical impacts of the multiple "prongs" of the whisk.

In the case of a soufflé, we use water vapor created during cooking to drive the enormous expansion in volume of a network of bubbles covered by unfolded proteins, as each gram of water becomes a liter of steam (a 1000-fold increase in space occupied!). Even if some of this steam escapes, we end up with a great increase in volume and a highly porous structure full of voids (see Figure 6.4), the walls of which are made of protein, which has developed a much more rigid solid structure because of the heat-induced denaturation.

In beer, proteins from the hops and malt used in the brewing process supply the foam-stabilizing agents. Foams in different kinds of beer, however, can differ in the gas that is used to produce the foam. In lagers, for example, carbon dioxide is used, which is able to dissolve in and move throughout water, allowing it to coalesce into large bubbles that disperse rapidly, whereas in Guinness nitrogen is used, which is not as soluble in water and tends to stay in its original bubbles as a result, and so the foamy head persists much longer and more dramatically if poured slowly, carefully and in stages (particularly thanks to the color contrast to the darker liquid below).

Fats and oils interfere with the surface chemistry of bubbles, however, and can result in unstable foams or foams that collapse. So, when making frothy milk for a coffee drink, low-fat milk will work better than full-fat milk. When we heat a high-fat product by bubbling steam through the milk, the fat liquefies at the high temperatures reached and spreads on the surface of the bubbles, destabilizing them by interfering with the delicate water–gas interface and the position of stabilizing molecules on this surface. In a lower-fat system, the key impact of the heat will be to unfold and denature proteins, giving a more solid and stable coating on our bubble surfaces. In fact, if we foam cold milk with air, air is more soluble at lower temperatures and so foaming works well, while at higher temperatures the liquid is less viscous and will drain out of the spaces between bubbles more easily, also contributing to the less-stable foam. So, comparing foaming of milk at low and high temperatures involves a trade-off between the positive effects of protein stabilization and the increased tendency for drainage and collapse at higher temperatures.

Some images of foams in food are shown in Figures 6.4 and 6.5, the first showing a light and expanded caramel mousse, in which whipped eggs created the light and fluffy structure, while later high-temperature cooking (or sometimes the judicious and careful use of, ahem, a blowtorch) induced Maillard reactions at the surface to give a brown and crispy surface effect. Studying the inner structure of a foam (Figure 6.5 shows a meringue) reveals a structure as porous as that of Styrofoam, with remarkably symmetrical bubbles of different sizes taking up the majority of the volume of the material, ready to collapse once the mechanical forces in the mouth are applied.

On the subject of consumption of these kinds of products, we don't want our foams to be too stable, as that would give a peculiar mouthfeel and make products hard to pour or otherwise work with,[15] so the key is to be able to have a foam for exactly as long as we need it, but no longer.

Chocolate

Another famous food product that depends for much of its physical and sensory properties on the properties of fats and oils is chocolate.

Chocolate is known to have arisen a very long time ago in the Americas, and the word itself seems to owe its origins to the Aztecs, who apparently enjoyed the bitter treat (the word's roots relate to the Aztec word for bitter) in between human sacrifices. Columbus was perhaps understandably impressed with the discovery of this treat during his voyages of exploration (probably more impressed than he might have been with an offer of sacrifice) and brought it back to Spain, from where it rapidly conquered Europe with a speed and ease that would have impressed Napoleon or Genghis Khan; an early improvement in terms of maximizing acceptance for

---

[15] For example, in the popular children's treat called the ice cream float, bubbles from the fizzy drink become super-reinforced with proteins from the ice cream dropped into it, and become hyper-stabilized, giving a very thick foam that almost needs to be eaten with a spoon.

some consumers was the inclusion of milk in the recipe to give milk chocolate, in which the bitterness was toned down somewhat.

The key source of the materials that make up the heart of chocolate is the cocoa tree, from which seed pods can be recovered. In traditional processes, these are allowed to sit in the sun's heat to increase their temperature and thereby kick off a set of biological and chemical reactions that result in the death of the seeds, the breakdown of their cell walls, and a number of reactions that ultimately lead to the production of key flavor compounds.

These fermented beans can then be roasted to develop rich flavor and color compounds (through the Maillard reaction, which we will come back to many times because of its omnipresence in changes on heating food), and then cracked open to remove their kernels, which are ground to a thick oily liquid, called cocoa butter, in which float particles of cocoa powder. This mixture can be milled and refined to break down the particles therein, the smaller the size obtained here giving a smoother and less-coarse final product. Add sugar and milk powder at this stage and milk chocolate will result, after a process called conching, in which the mix is held at warm temperatures while being worked back and forth by a roller. During this step, some water and volatile compounds are removed and the sugar and other solids worked into a smooth well-mixed system that can be poured into suitably shaped molds and cooled, solidifying as the cocoa butter molecules pass the temperature at which they prefer to be solid rather than liquid, and a crystal network results.

Part of the delight of chocolate then results from the reversal of this process, when the chocolate is exposed to the warmth of the mouth, and a slow melting occurs as the cocoa crystals revert to their liquid form, just in time to be swallowed!

The behavior of crystals in chocolate can also cause some undesirable problems, such as what is called bloom, when there are white patches seen on the surface of an otherwise nice-looking bar. This is down to the fact that we need to have the right kind of crystal present, meaning lots of very small crystals of the right type[16] that behave themselves.

If we subject the product to undesirable fluctuations in temperature, where it warms and almost melts and then cools and solidifies again, we can form unstable crystal shapes and networks, which basically don't coexist well with the other solid materials in the chocolate, such as sugar and cocoa particles. These can be rudely shoved out of the way by the unruly and chaotic behavior of the fat crystals and end up pushed to the outside, where they gather and eventually are concentrated enough to give a visible white patch.

Overall, when we consider the properties of fat in food other than those that could cause nutritional concerns (not to dodge the bigger questions, of course!), we see

---

[16] The fat in chocolate can actually form multiple different crystal forms, of different sizes, shapes, and consequences for product characteristics, and the secret to successful manufacture or use in recipes involving steps such as tempering is controlling the exact forms of crystals present.

that, were fat to disappear suddenly (by magic or legislation, whichever works best), we would lose something important from food, something that, despite its bad reputation, confers texture, richness, and a multiplicity of other desirable sensory characteristics. When fat is present, something is brought to food, and its absence leaves voids beyond the structural, for better or worse.

7

The Many Roles of Microorganisms
THE GOOD, THE BAD, AND THE UGLY

The diversity of microbial life in food

As mentioned already several times, the world of living things can be divided quite simply into that which we can see (animals, plants, us!) and that which we cannot see. We share this planet with microscopic life that actually far outnumber the life-forms we can see and whose importance to our lives across a huge range of areas is completely out of proportion to their size. We need the assistance of microscopes or other tools to reveal the incredible diversity, richness, and sheer vastness of this hidden world.

In terms of the sphere of human life with which we are concerned in this book, food, we worry about two things to do with microorganisms in food, which are safety and spoilage, but these are not the same thing. For example, milk containing a bacterium called *Pseudomonas* could look green, be stinky, and have lumps floating in it, but could be quite safe, while milk containing listeria could look fresh as could be but would make you very ill, perhaps even fatally, were you to drink it. In addition, yogurt containing bacteria called Bifidobacteria might not only be neither unpleasant nor dangerous but might actually be good for you, as these are probiotic bacteria, which are believed to colonize the human gut and help keep us healthy.

So (cue Ennio Morricone music, and distinctive whistling), bacteria in food can be good (like the probiotics), bad (like the pathogens), or ugly (like the types that cause spoilage). Of course, the population of living things we cannot see in food is much broader than bacteria too and encompasses viruses (generally these don't come in good or ugly variants, and are usually simply bad news, as when they cause food poisoning because of their contamination of products like oysters) and fungi such as yeasts and molds.

Some yeasts and molds are "good," in that they are crucial for the production of beer, bread, and many other products, while others fall into the ugly camp (with occasional cases of badness, when certain types produce toxins called mycotoxins), and are responsible for spoilage of fruit, for example, because of their ability to grow under conditions that are so acidic that most other microorganisms would struggle to survive, let alone thrive. Yeasts and molds also illustrate our cultural (in the social sense) selectivity when it comes to these kinds of culture (in the microbial sense); put blue mold in Stilton cheese and it is an essential element of the quintessence of the product, but put the same mold, or a near cousin, on the outside of an orange and disgust and likely discarding will follow.

Similarly, when wild-type bacteria that are found rampant in nature land in an opened milk container and cause it to ferment before we apply the milk to our cornflakes, this is bad, yet thousands of years ago we tamed essentially the same kinds of bacteria and routinely now use them to make yogurt (which we also learned long ago not to put on our cornflakes).

So, microbial life, as it applies to food, can be divided into the wild, untamed, and undesirable (for reasons of safety or food quality) and those that we have put to our service as a key partner in the production of certain food products (fermented foods), again long before the unseen agent responsible was identified or even dreamed of. This is just like the divisions between plants and animals that we discussed in Chapter 2.

Clearly, food and microorganisms come hand in hand, and the science of food preparation and preservation depends on a profound understanding of both. For centuries those who have handled and processed food have applied instinctive learned experience of how to control microorganisms before the term was even dreamt of, through pickling, fermentation, cooking, cooling, and myriad other steps. For the last century, though, we have increasingly understood the science of microbiology, and now can look at all that was done and known to work and say "ah, so that's why . . . ."

Even today, as more advanced techniques based on molecular biology are being applied to characterize food, scientists are almost daily discovering new species, new relationships, and new systems that show in ever-more-dramatic fashion what a complex, teeming world exists below the limits of resolution of the human eye. For example, in a study published in 2014 in the journal *Cell,* scientists from Harvard University analyzed 137 cheese samples using state-of-the-art technologies and found such surprises as bacteria commonly found in the sea being isolated from cheese made nowhere near an ocean, leading to the possibility of their hitching a ride on salt added during the cheese manufacture and traveling huge distances in this way.

There are very few food products that are completely free of microorganisms, and these are those that have been heated to extremely high temperatures, such as ultra-high-temperature (UHT) treated milk, or canned foods, which are so sterile that we can store them outside the fridge without fear of anything growing in them. Dry foods also might not be exactly sterile, but are so arid that microbial life, which is moisture dependent, cannot thrive.

Effectively, however, every other food system is quite simply a zoo, where it is quite likely that we haven't even properly characterized all the exhibits. What keeps us safe are the properties of food, the processes we apply, the ability of our bodies to fight and kill bacteria on ingestion, and often simple luck.

So, throughout history, we have learned to tame safe microbes and make them do our bidding while also learning huge amounts about the unsafe ones and how to kill them or keep them out of our food, for example, through the addition or formation through fermentation of chemical preservatives such as acid and alcohol. Food scientists need to understand how to control bacteria, tame them, use them for our own purposes where they fit these, and very frequently kill them when they don't. To do this, food scientists need to learn everything they can about their enemy so as to exploit their weaknesses and learn how to either destroy them or make their environment so inhospitable that they give up of their own accord.

Life beyond our eyes: The biology of microorganisms

Before considering how we harness, control, avoid, and outright seek to annihilate the multitudinous forms of microbial life we find in our food or the raw materials we seek to convert into food, it is worth considering what types of life we are actually talking about.

All microorganisms are at heart a single cell, in a dramatic demonstration of the minimum conditions that life requires to exert itself.

The human body, by comparison, contains a vast number of different types of cells, with specialized functions, from our skin to our blood, brain, and organs, each of which shares certain core elements and structures, but that have been specialized so that they serve different and critically unique functions.

In contrast, microorganisms in most cases make do with just one cell, which must contain everything needed for life and then be able to multiply by division to yield more cells, which become new individuals. While body cells in an animal or plant are in most cases surrounded by other cells or friendly parts of the organism, each microbial cell is an island, and must live independently within the environment in which it finds itself. From this viewpoint, it is perhaps even more admirable that bacterial cells can be so successful in colonizing and adapting to almost every possible environment in which they have found themselves, from the hottest to the coldest, across extremes of acidity, saltiness, aridity, and the presence of substances most other organisms would find to be highly toxic.

Bacterial cells thus must differ from those we find in animals or plants (multicellular organisms) in a number of ways that reflect their need for independent living, such as being surrounded not just by a membrane but also, in many cases, by a stronger cell wall as well, reflecting their existence in an environment in which everything outside the cell is different and potentially hostile.

When we refer to bacterial cells as microorganisms, we reflect the fact that they are really really small. But how small is small? First, the answer to this question depends on what shape the cells are in, as most bacteria are either found in spherical shapes (called cocci) or oblong (sausage) shapes called rods. In either case, they are around two-tenths of a micrometer in diameter, and maybe 10–50 times this in length (for the rods). To understand these units, first think of a millimeter, which is about the thickness of a credit or bank card, and then divide that by a thousand, to get a micrometer. That is very small indeed. The period at the end of the previous sentence is around one-tenth to half a millimeter in diameter, so bacteria might be one-hundredth of that size; to put that another way, we could fit several thousand bacteria onto something the size of that period. On this basis, the size of each individual cell is clearly below the limit of resolution of the human eye, and so we cannot see them as individuals, and for millennia remained blissfully unaware of the unseen zoo with which we shared our world.

Animal cells, by comparison, are around 10 times the size of bacterial cells, and plant cells are larger still. So, not only do bacterial cells need to pack all the requirements for life (albeit pretty simple life) into one unit, as opposed to dividing functions between different types of cell, but they have to deliver this in an extremely small unit. It is hard not to admire the success and efficiency of these forms of life, until we remember the problems they can cause us.

Bacteria are genuinely ubiquitous in our world; if we could imagine that each bacterial call, whatever species it belonged to, suddenly started to shine brightly like a little lightbulb, so that their presence could be seen visually, we would be constantly dazzled as almost every surface and item around us would shine brightly, revealing the fact that they are all covered in bacteria.

The world of bacteria and our world intersect in a bewildering number of ways. Some are good for us, some are bad for us, and some are haughtily indifferent to us. Some make us sick, some make us cheese, and some make us retch at the sight of a putrefied container of milk left too long in the fridge.

Images of some common bacteria of relevance to the world of dairy products are shown in Figure 7.1. These images were taken with a relatively low-power light microscope, following staining to visualize the cells. It can immediately be seen that there are two types of cells, the round cocci (*Staphylococcus* and *Lactococcus*) and the rod-shaped bacteria (*Bacillus* and *Lactobacillus*), but otherwise they look reasonably similar and rather innocuous. If you had taken the sample of culture from a dairy product, however, you would be quite concerned to find the top two unwelcome guests, while the bottom two would be very normal to find in cheese.

Thus, in a food context, we rely on bacteria and the reactions that they promote for the very existence of many food products, while other species are of interest primarily in terms of understanding better how to control or kill them.

Bacteria are not even the simplest forms of life, as in the 1890s it was discovered that liquid filtered through a material that should block the passage of something the size of a bacterial cell could remain infectious, and hence there is something (nasty) that is significantly smaller than bacteria, which are viruses.

Viruses have such simple structures and components that they basically sit uncomfortably on the divide between what we regard as alive and what we regard as not alive, but yet can infect cells (plant, animal, or even bacterial), transmit their genetic information, and turn their host into a machine for making copies of themselves, thus fulfilling the basic expectation of life that, at a minimum, it reproduces. Simple they may be, yet viruses such as norovirus can cause severe gastroenteritis and, as the rate of illnesses that are due to bacteria in food has gradually fallen over the years, that which is due to viruses has risen, and it seems likely that increasing focus will fall on these deceptively basic forms of life, once described as "a piece of bad news wrapped up in protein."[1]

Also cohabiting our world yet lurking beyond our ability to see are fungi, which include yeasts and molds. Indeed, one of the most common microorganisms that mankind has pressed into service over the centuries must be yeast, which we find in products from wine and beer to bread. The reason yeasts have proven so useful for us is that they are highly efficient biological machines for turning sugar into carbon dioxide and alcohol. When we make bread, we use types of yeast that produce very little alcohol, whereas the more fermentative strains are of course what is of interest to the brewer. As we will see in Chapter 8, the activities of yeast are also dependent on the conditions in which they find themselves, and whether they make alcohol, and how much, can depend on factors such as the availability of oxygen, the types of sugar present in the raw materials, and the temperature at which they grow.

Turning understanding into a weapon

Every bacterium has a defined range of tolerances for conditions in which it can survive, and all have their weak spots, be it a certain acidity (pH) below which they cannot grow, or a certain level of salt, alcohol, or sugar above which the conditions simply become so severe that survival is not an option. Temperature is also a key governing consideration, with most microorganisms being so-called mesophiles that prefer to grow at body temperatures (around 37 °C), but some being quite comfortable in more chilly refrigeration temperatures (so-called psychrophiles), while others can tolerate (thermoduric) or even enjoy (thermophilic) high temperatures.

The astonishing diversity of the microbial world means that bacteria have been found in even the most bizarrely hostile conditions (far more extreme than found in food), such as the salty soup of the Dead Sea (halophiles, who laugh at levels of salt that would make most bacteria weep out their contents and shrivel up), or even boiling conditions at extremely alkaline or acidic pH values, as found in hot springs on land or deep undersea. So, some of these demons actually live in salt, while others are essentially fireproof, and so the trouble they can cause can find you almost anywhere.

Thankfully, in food, such extremely weird species are not found, and, as in so many areas of food science, we worked out a long time ago how to use the weaknesses of more typical microorganisms against them, by adding salt when we

---

[1] By the great British biologist Sir Peter Medawar.

brine, increasing alcohol levels when we ferment, or lowering pH when we pickle in vinegar.

The most cunning trick that bacteria can pull that causes problems for food processors is that practiced by those species that can resist extremely high temperatures, laughing callously at the inability of traditional processes such as pasteurization to even tickle them. They manage this trick through their ability to retreat into a supertough state called spores, in which they slow down their metabolism, secrete a tough coat around themselves, and hunker down to wait out the fiery inferno we apply when we heat the food, biding their time until conditions are more temperate and it seems safe for them to come out of their shell (germinate) and wreak their merry havoc.

So, when we heat milk or other products to temperatures sufficient to pasteurize them (around 70–80 °C), we generally kill the non-spore-forming pathogenic bacteria and most of the spoilage bacteria, but do little to the spores; for this reason, we must apply another trick to keep them under control, such as keeping the product in the fridge. This works because most spores don't like such low temperatures and won't wake up from their hibernation. So milk stays safe when kept in the fridge (take it out and leave it warm up, though, and all bets are off), and when we drink the milk the spores are killed by our stomach acids. The surviving spoilage bacteria that are not spore-formers, and sometimes some spore-formers that can tolerate low temperatures, grow slowly and eventually cause the milk to go off.

To make products that don't have to be kept in the fridge, we need much higher temperatures (above around 110 °C), such as are applied when we heat canned foods in a retort for a few minutes, or at much higher temperatures for a few seconds when we apply UHT processes. These are the conditions we need to reach where we can finally genuinely kill the spores. However, such temperatures are highly likely to damage the quality of the food in terms of nutritional or sensory damage, so we need to understand exactly what the minimum conditions needed to kill the spore are, and so we need to know a lot about how spores respond to heat.

Rather than focus entirely on the negative side of bacteria in food, let's first spend some time celebrating their successes, through some examples of how fermentations deliver some of our most celebrated food products.

The happy marriage at the heart of yogurt

We have discussed the principle of how yogurt is produced in Chapter 3, when we discussed the propensity of casein in milk to form complex three-dimensional structures.

The key to production of yogurt, as for many fine things in life, is teamwork, in this case at the microscopic scale. Production of classical yogurt requires the joint and cooperative efforts of two different species of bacteria, one called *Streptococcus thermophilus* and one going by the particularly fancy name of *Lactobacillus delbrueckii* subspecies *bulgaricus*.[2] Either of these bacteria will just about grow in milk, but one imagines that for them to do so would be an incomplete existence, from which something was inexplicably missing, as it is when they are in each other's company that the real magic happens. I once read an Italian scientific article that referred to the "marriage" of the bacteria in yogurt, and indeed a successful yogurt culture is like a successful marriage, with each species supporting the growth of development of the other in a true partnership, and where the whole is definitely greater than the sum of the parts.

The same article showed beautiful photographs taken down a very powerful microscope of the delicate lacy filigrees of protein that make up the yogurt gel and, within this gossamer network, lots of tiny cavities in which were found clusters of the bacteria, the distinctive round cells of the Streptococci and the longer sausage-shaped cells of Lactobacilli nestled together happily. At this scale, far beyond the ability of the eye to see, yogurt is a vibrant lively thriving community, with two populations from different cultures integrated so successfully as to represent a single happy ecosystem.[3]

This can be seen in the micrograph in Figure 7.2, in which the casein network can be seen, within which are thriving many microbial cells in close proximity, from clusters of the ovoid Streptococci to the long rods of Lactobacilli. In scientific terms, the success of this relationship is based on an exchange of much-needed nutrients between the species, with each producing during their growth in milk a by-product that helps the growth of the other.

To produce yogurt at home is quite simple, exploiting the fact that yogurt is a live system. Taking some fresh milk and adding a small amount of yogurt will seed the milk by introducing the culture to a new blank sheet, as it were, of untapped nutrients and growth potential. If the mix is then held at a suitably warm temperature (Streptococci in particular like it quite toasty, hence the "thermo" part of their name) it will gradually ferment and gel, and some of the resulting yogurt can be taken and used to inoculate a fresh batch of milk, and so forth.

In industry, the bacteria will be supplied in frozen or dried form as a highly concentrated culture, effectively of sleeping beauties, ready to be woken up not by a kiss but by a gradual warming up in a batch of suitable milk. After such enforced slumber, one can imagine the bacteria will be hungry, and the presence of bountiful nutrients will stir (or shake) them into rapid action, as they gorge on the milk sugar, lactose, and convert it to lactic acid. This reduces the pH gradually (over a couple of hours) to the point where the casein (as discussed in Chapter 3) reaches its isoelectric point and forms the three-dimensional network at a microscopic level that, at a macroscopic level, magically transforms the milk from liquid to semisolid gel.

Ideally, yogurt should be a "perfect gel," smooth and even, ready to scoop and savor, but one of the key undesirable defects in yogurt is what is called syneresis,

---

[2] Interestingly, this was announced in 2012 as the national microbe of India.
[3] Many countries could learn a lot from considering the lessons of yogurt.

where the gel has contracted and some loose liquid (whey) has escaped and floated to the surface of the gel, thereby becoming the first thing consumers see on opening their pots of yogurt. The avoidance of syneresis is in fact one of the key drivers in the technology of how yogurt is processed today, and many of the steps applied to the milk before the bacteria are added and do their wonderful thing are designed specifically to avoid this defect from causing unnecessary mild distress to consumers.[4]

Today, two main types of yogurt products are produced, sometimes called stirred and set.[5] These differ, as we will see, in how exactly they are fermented and the physical treatment they receive during and after fermentation; they can readily be differentiated in practice, stirred being much more fluid than the more gelled and solid set variety.[6] In either case, the gel must not show undesirable syneresis.

If raw milk, or even normally pasteurized milk, was used as the basis for making yogurt, a pretty weak gel that was very prone to syneresis would result, and it wouldn't resemble a good modern yogurt at all. So, what the bacteria actually ferment is instead a carefully formulated and processed fortified milk-based mixture, precision-engineered to give the optimal texture.

In terms of process, the objective is to get every single part of the milk that could usefully play a role in generating a solid structure to do so. The core structure is based on the casein network that has formed, as discussed in Chapter 3, but additional structure and strengthening links in those casein chains can be added by denaturing the whey protein fraction in milk,[7] causing these highly reactive proteins to unfold and link with the casein, adding as it were a reinforcing layer of scaffolding around the casein network as it forms, and a separate and secondary but interlinked gel structure. This is achieved by heating the milk to temperatures around 80–90 °C for several minutes, conditions far above those used for typical pasteurization of milk. Such heating also has the bonus effect of more or less sterilizing the milk and removing any possible competition for our culture once added, or unwanted bacteria that might grow when we warm and hold the milk for our yogurt fermentation.

In addition, a further element of the milk that can help build structure is the fat globule discussed in Chapter 6, which we saw can be induced to interact with the proteins in milk when it is homogenized, and so the yogurt mix is homogenized to get this interaction to take place. In this case, when the casein reaches the pH value at which it gels, the casein-encrusted fat globules will line up as further bricks in the network wall, adding yet more strength and resistance to syneresis in the gel.[8]

---

[4] Possibly a classic example of a "first-world problem"!

[5] A third variant, the James Bond, is shaken but not stirred, but it has never taken off.

[6] A highly scientific test for differentiating them involves holding an open pot over your head, or that of a willing volunteer. If it doesn't land on your or the volunteer's head, it is set yogurt and, if it does, don't blame me.

[7] Causing them to unfold from their original structure, with the result that, in their unfolded state, they tend to become very reactive and interact with (stick to) other proteins.

[8] As discussed elsewhere, syneresis is a key part of making cheese, as we want to expel whey during its manufacture; basically, everything we do in yogurt to avoid syneresis we reverse in cheese, and so

In terms of formulation, producers often also exploit the fact that the gel network is made of protein, both casein and whey protein, by adding more protein, in the form of milk powder or more expensive purified milk protein fractions. In addition, as the main objective in avoiding syneresis is to control the mobility of water, yogurt formulations often include the class of molecules we met in Chapter 5 that have the best water-controlling properties of all, which are polysaccharide gums and stabilizers. These hydrophilic sponges are extremely useful in yogurt, through absorbing loose moisture in the structure and stopping it from flowing out where it shouldn't. A common stabilizer added is pectin that, because of its fruity origins, is quite comfortably at home in the acidic conditions of yogurt.

So, before bacteria get anywhere near the fermentation, the formulation and process are optimized to make the gel formation process work as well as possible in terms of gel structure.

At this point, it is also common to add any flavors, sweeteners, colors, and other additives that mightn't be part of the structure agenda but give consumer-friendly flavors and characteristics. It is worth noting though, and this is a point relevant to many food processes, that adding such ingredients always involves a lot of considerations. Questions to be asked in this regard (besides the source and level of addition of the additives) will include the resistance of the ingredients to heat and physical processes such as homogenization (for example, might tasty chocolate pieces bung up the fine tubes in that machine?). Also, if the rather severe heat treatment applied might change the flavor or color, it may be preferable to add that ingredient after heating, but of course this then raises the possible problem of microbial contamination *after* the key step that makes the product safe and stable, which needs to be considered. In a further consideration, some fruits might interfere with the added starter culture because of the presence of natural inhibitors, and need to be added after fermentation (if possible), or else perhaps separated entirely from the mix, for the consumers to add themselves later. This is easily achieved in the kind of segregated packs that have become common in many markets in recent years. This is also an easy way of keeping mechanically sensitive particulates like chocolate or cereal pieces away from machines that might otherwise mangle them, or vice versa.

Then, having appropriately formulated and pretreated the mix, it is time for the starter culture to be unleashed and start the biological part of the process. At this point, however, there is a further key decision to be made, which is whether mix plus starter is added directly into pots or the fermentation takes place in some sort of bulk tank, with the resulting yogurt being packaged later. This is where the difference between set and stirred yogurt mainly comes in, as the former is fermented in the final package, with the lid being put on after filling and not removed until a hungry consumer is ready to eat. In the case of stirred yogurt, on the other hand,

---

milk for cheesemaking would never be homogenized or heat-treated as severely as that used for yogurt, as the goals and desired gel properties are completely opposite. Yogurt is the uncheese, and cheese is noghurt.

the product is packaged after the fermentation, and this mechanical movement and handling is part of what gives that product its more fluid nature. For stirred yogurt, there is also the option to add final ingredients (maybe including the fruit that would have interfered with fermentation) at the end before the lid goes on.

The fermentation is then allowed to progress under balmy warm conditions to the point where the bacteria have produced enough acid from their lactose-slurping endeavors to bring the pH to around 4.6, at which the casein–whey protein–fat complex has gelled, hopefully into a nice non-syneresing structure. In parallel, other metabolic pathways have produced small molecules such as diacetyl and acetaldehyde, which contribute even at low levels to the distinctive flavor and aroma of yogurt.

When the pH arrives at its intended level, which typically takes 3–4 hours of undisturbed fermentation, the bacteria can be relieved of duty either gently, by cooling the yogurt and retaining a live product with dormant bacterial cells, or more harshly and arguably ungratefully by killing off the cells using heat treatment. The latter gives a longer-life product and can be combined with homogenization after gelation to give more liquid drinking yogurts.

So, next time you open a pot of yogurt, and see a smooth unseparated creamy mass within, you can appreciate a happy marriage, not just of the bacteria within but of microbiology and the technologies of food processing and formulation.

Kefir and probiotics: Are bacteria, alcohol and bubbles the secret to long and healthy life?

There are a wide range of variants of yogurts available, based on the happy fact that there are many species of bacteria and other microorganisms that have the beneficial combination of properties that they (A) grow well in milk, (B) produce interesting compounds like lactic acid and desirable flavors, and (C) are perfectly safe for human consumption. Many countries around the world have their own locally fermented milk drinks or products, including labneh, kishk, lactic drinks, fermented buttermilk, and many more.

One particularly interesting product, called kefir, has an almost legendary status for its reputed ability to extend the life spans of its consumers, a reputation based on studies over a century ago of long life expectancy of inhabitants of the Caucasus Mountains. This work was undertaken by a Russian scientist called Élie Metchnikoff[9] who reportedly spent years in remote villages around the Caucasus Mountains, observing what the local Balkan peasants were consuming, and coming to attribute their uncommonly long life span to the near-daily consumption of kefir and fermented drinks containing a bacterium then called *Lactobacillus bulgaricus*.[10]

---

[9] Metchnikoff won a Nobel Prize in 1908 for his studies of how cells in our bodies fight infection, and in advancing years became interested in why certain populations maintained healthy vigorous lifestyles, with less incidence of senility than others, which led him to study kefir.

[10] The name of which is suggestive of its identification in studies from Eastern Europe.

Some of the products he encountered seemed to be yogurt-like (called koumiss, sometimes made from horses' milk) and others more carbonated and almost fizzy—these latter drinks are kefir. The mechanism he believed responsible involved inhibition of growth of bacteria that generated toxins in the human digestive tract, which could, if unchecked, limit life spans.

A unique property of kefir is the nature of the culture that ferments it, which comes not in any traditional liquid or dried form, but rather looks like small white lumps of material that resemble popped popcorn. These so-called kefir grains are composed of a complex matrix of protein and polysaccharide within which are embedded a thriving and frequently extremely complex microbial ecosystem,[11] the main attractions of which are milk-fermenting lactic acid bacteria and yeasts. While the presence of the former kind of bacteria mean that kefir fermentation results in a decrease in pH and yogurt-like gelation, the latter are microorganisms far less commonly associated with yogurt. Yeast, when present, operates a whole different set of metabolic pathways for the utilization of milk components than bacteria, and ends up with very different products. Specifically, in terms of the character of kefir, they can produce alcohol and carbon dioxide (exactly the same products of yeast-driven fermentation we will discuss later as being at the heart of the production of products such as beer and wine). In turn, as they work in the milk, the grains grow and swell and can be recovered after fermentation, washed, and used for another fresh batch of milk.

The microstructure of a kefir grain, imaged with an electron microscope, is shown in Figure 7.3. The most basic level of structure here is the small clusters of coagulated material (mostly milk protein) seen in knots and clumps throughout the the image, while the large cells are the yeast and the long chains of rods are the bacteria that convert the lactose to lactic acid.

In some areas of the world, including rural Ireland, such grains are found in many kitchens, sometimes known as buttermilk grains, and are used to keep a batch of beverage in fermentation on an ongoing basis.

While the levels of alcohol and carbon dioxide generated in kefir are normally not enough to give what would be considered an alcoholic fizzy beverage, careful management and encouragement of the grains, and the addition of sugar, can lead to a product that has been referred to as "milk champagne," which is really quite pleasant. While such products may not be scientifically guaranteed to extend your life span, they could certainly make the life span you have more interesting!

Going back to the other of those healthy products first publicized by Metchnikoff, in recent years, an increasingly busy section of the chilled dairy part of stores and markets has involved products (found much more easily than kefir in most places)

---

[11] Some of the other bacteria found in kefir have probiotic properties, and one was shown some years ago by Irish scientists to produce a very potent molecule, called a bacteriocin, that can inhibit the growth of undesirable bacteria in a form of interbacterial chemical warfare. The presence of such species is likely to account for at least part of reported health benefits of kefir.

whose labels contain the word "probiotic," and we frequently hear about the idea that such products contain "something" that will make our innards a happier place.

I remember an ad that used to play on Irish radio describing one such product as containing a certain number of billions of probiotic nutrients. Wow, one might think, that ingredient list must be really really long! This was a clear example of an apparent fear of mentioning the word "bacteria" in the context of what was supposed to be something good to do with food for fear of confusing consumers.

Since Metchnikoff's first recognition of the health-promoting role of probiotics, almost a century of studies has explored how bacteria that grow (mostly, but not only) in milk can, if consumed, colonize our innards and render them less susceptible to growth of unwanted kinds of bacteria, and improve our health as a result. Probiotics are defined by the Food and Agriculture Organization of the World Health Organization as "live microorganisms which, when administered in adequate amounts, confer a health benefit on the host."

The idea behind how probiotics work is that our digestive systems already play host to a huge number of bacterial species and numbers. Many of these are harmless or naturally beneficial, as introduced in Chapter 2, but illness can result from the entry and growth of unwanted harmful species. Regular consumption of probiotic bacteria (including members of some families of bacteria we have encountered in cheese and yogurt like Lactobacilli and others such as Bifidobacteria) should gradually drive the populations toward a dominance of beneficial species, who then undertake a sort of surveillance role, making the locality far less tolerant to undesirable invaders and perhaps producing beneficial molecules in the process.

For this to be effective, large numbers of probiotic bacteria need to be consumed, at least 1,000,000,000 cells per day. If these were people, that would be one-eighth of the world's population, but, as bacterial cells are very small indeed, that amount of them can fit into a small shot-sized plastic bottle, as probiotic products are often packaged in.

Effects attributed to the regular consumption of probiotic products include stimulation of the immune system, protection against inflammation and cancer, alleviation of constipation, and treatment of inflammatory bowel disorder.

In recent years, there has been a lot of intense scrutiny of whether probiotic products can be directly linked to specific health benefits, to an extent that a direct cause-and-effect linkage can be concluded to exist. As a result, in Europe today there is significant caution about making overly specific claims on the labels of probiotic products, unless expensive clinical trials have been undertaken that can provide the direct evidence of the bodily changes resulting from probiotic consumption.

In the meantime, research continues to be published reporting a whole range of benefits for the consumption of probiotics. At present, a particularly strong theme appears to relate to the relationship between the bacterial population of our guts and our emotional state, including conditions such as stress, anxiety, and depression (a whole new take on "gut feelings"). The mechanism by which such a relationship could exist concerns nerve connections between our gastrointestinal tract (GIT) and the brain, and the ability of beneficial bacteria to affect signals traveling along this axis, feeding positive news into this internal communication superhighway, as it were. Bacteria in the GIT could also modify the production of key neurotransmitters and hormones, which would also help to influence our mood.

Bacteria are only part of this story, though, as it is also increasingly clear that diet influences these communications, both through consumption of molecules that directly impact on factors such as cognition (such as specific fatty acids and proteins) and through food components known as prebiotics that foster the growth of beneficial bacteria. Examples of prebiotics are specific complex sugars (such as inulin, a compound found widely in vegetables such as onions, garlic, asparagus, and beets, as well as bananas and wheat) that encourage the growth of probiotic bacteria.

There is no doubt that the coming years will be very interesting in the field of probiotics, in terms of whether promising results from laboratory studies can be translated into sufficiently strong clinical evidence of direct health benefits, and then it can be really determined whether our microscopic life companions can be fully tamed to make them partners in our internal and external health.

Fermentation and cheese

In Chapter 4, we discussed how protein breakdown takes place and can be examined and evaluated in cheese, and how such a breakdown is a key step in how the texture of cheese develops during ripening and how differences in levels and type of protein breakdown can be the reason one variety of cheese differs from another.

The agents (enzymes) responsible for this breakdown originate mainly from the milk itself and the coagulant (rennet) we have added to turn the milk to the gel that forms the basis for the cheese to follow.

However, beside this coagulant being added to initiate the conversion of milk to cheese, we always add a starter culture, for cheese is a fermented food, and, while the enzymes from milk and coagulant are responsible for building (and then changing) the structure of the cheese, like builders, the microbes are really the interior decorators who come along and fill the interior of that structure with flavor, character, and color. They bring life in every sense to the heart of cheese, and the difference between varieties can be due to careful selection of exactly the right decorators for the task.

Just as protein comes to cheese from milk, so do lipids and some of the sugar, lactose, in the milk, and these are also the raw materials on which bacteria will work their magic. Some of these bacteria were added by the cheesemaker, while others are recognized to arrive completely accidentally, from the air and environment of each cheese factory; they are such regular parts of cheese ripening that they have their own name (nonstarter lactic acid bacteria, or NSLAB, pronounced as if there is an "i" between the N and the S). Indeed, in mature cheese, because the death of the added starter in the early stages of ripening, such NSLAB typically represent the main bacteria found.

The exact pathway by which bacteria in the cheese handle the small amount of lactose present in the early weeks of ripening can be a defining characteristic of varieties such as those associated with Switzerland, like Emmental with its large eyes. In most types of cheese, ripening takes place at relatively cool temperatures, like 8–15 °C. In the case of Emmental, however, the development of eyes requires a particular lactose-based reaction to take place, leading to the production of carbon dioxide, which accumulates in large holes created in the cheese at random sites where imperfections in the original curd allow gas to accumulate;[12] the bacteria that catalyze this reaction grow best at warm temperatures, so the ripening of this cheese requires a "warm room step" in which young cheese is placed at around 25 °C for a few weeks, during which the eye development is initiated and the final cheese structure determined.[13]

Cheese ripening also directly affects the other components of the milk that have become entrapped in the curd. While the deconstruction of proteins was my focus in Chapter 4, the sugar (lactose) and lipids from the milk are also raw materials for a complex series of pathways that ultimately lead to development of particular flavors and cheese characteristics.

In fact, in cheese types with much stronger flavors than Cheddar, such as mold- or smear-ripened cheeses, the key to development of their piquant peculiarities is the fact that the breakdown products of milk fat frequently have a rather striking impact on our noses and mouths. Milk fat, when tasted in the form of butter or cream, is rather bland in flavor, invoking of course creaminess, butteriness, and general wholesome dairy character. Take these large, homely flavored fat molecules, though, and feed them through a sequence of enzymatic and chemical transformations, particularly brought about by microbes found in or added to these types of cheese, and the end products can have a very different flavor profile.

Types of molecules resulting from fat breakdown in cheese can even have surprising associations with other sensory encounters; for example, carboxylic acids, found in strong cheeses like Limburger, are also associated with the smell of socks that have undesirably captured the exertions of their wearer.[14] Other products of fat metabolism in cheese include esters, which have fruity flavors, alcohol, and compounds called lactones are also associated with the flavors of foods like peach and coconut.

So, many of the compounds produced ultimately from milk fat during ripening have their own unique (sometimes positive, sometimes negative) flavor

---

[12] Swiss cheese is, yes, carbonated, and so could be called fizzy cheese, albeit with larger bubbles (eyes) than found in carbonated drinks. It is perhaps not surprising that the term "fizzy cheese" has never caught on. As so often mentioned in this book, the properties of many food products are determined by unexpected science, and in this case both these drinks and Emmental are similarly dependent on fizzics.

[13] The successful completion of this step could be evaluated by holding a slice of Swiss cheese 45 centimeters away from a tester and counting the number of holes in the cheese while blocking one of the tester's eyes at a time. This is a standard type of eye test.

[14] This is a foot note.

characteristics, and if we could somehow subdivide the flavor of cheese into its constituent parts we might identify them, welcome them, or reject them.

This hypothetical approach to breaking up the flavor of cheese (or, for that matter, any food product) into its constituent individual elements is actually achievable in a laboratory modification of the technique of gas chromatography, which separates compounds from food in a stream of gas. This might involve simply allowing a sample of cheese to stand and breathe in a sealed container, such that the air or gas above the cheese becomes enriched in all those volatile flavor compounds that have the Wright-ian ability to become airborne, with their ultimate destination being the receptors in our nose. This mixture of aroma compounds can then be sampled and passed along a long thin tube, such that they become separated from each other on the basis of some property that determines their rate of movement through the tube, such as their size. They eventually pass a detector that records their presence and the key discriminating property that has determined their separation. In an unusual move, however, just past the detector the compounds can then exit the machine and, without further ado, shoot along a tube into the nose of a tester (an actual person) who records observations on the aroma of that molecule, so that compounds 1, 2, and 3 in the mixture, which are likely to equate to molecules X, Y, and Z, are determined as smelling "grassy," "sulfurous," and "cow rear-end emission," respectively. This technique, gas chromatography olfactometry (GCO), is an example of where human and machine merge, combining artificial and actual intelligence in a cyberpunk fusion of flavor analysis.[15]

By such means the aroma of cheese can be divided into its constituent responsible contributors, but probably none of these individual notes will be found to smell like "cheese." In the early decades of cheese flavor research, it was anticipated that eventually, as methods for extracting and analyzing the compounds from something like ripe Cheddar got better, someday someone would have a Eureka moment, look at their cheese under analysis, and say "you reek as a direct and inarguable result of the presence of this particular molecule." This molecule could then be purified or synthesized and, if tasted in isolation, would immediately taste of its parent cheese, on which it had conferred its flavor.

It was gradually realized and accepted that life wasn't that simple, and there was no single compound responsible for Cheddarness, Brieness, or Parmesaness, no matter how Caerphilly the scientists searched. Rather, cheese flavor is the combined result of the simultaneous detection by nose and mouth of a molecular cheese board of many different compounds, and differences between varieties came down to which ones were present, depending on how the cheese was made and what unseen agents had diligently digested the simple raw materials supplied by the milk and their relative levels and interactions.

Later in Chapter 17, we will discuss the idea of note-by-note cuisine, where dishes can be prepared from a series of individual compounds that each serve

---

[15] Perhaps this could be called (flavor) Determinator.

specific functions in terms of flavor, aroma, or texture. The key idea is that food flavor is like a complex piece of music, which works only because all the individual notes have combined synergistically into something the collective impact of which is so much greater than the sum of the individual parts. The sound of a bassoon, tympani, or French horn individually may be distinctive and powerful, but put them together with many other sources of beautiful sound in an orchestra, and the combination can be simply magical. Cheese in this analogy is the ultimate symphony of note-by-note flavor construction, where varietal differences arise like different pieces of classical music, combining basic elements (molecules or instruments) in different ways, at different intensities, to deliver a combined impact of complex beauty.

The culture wars: Battles on a scale we cannot fathom

Sometimes, a cheesemaker ends up with a mysteriously "dead" vat of cheese, where the culture fails to thrive and the expected drop in pH that is due to the lactic acid bacteria's fermentative frolics fails to materialize. This might well be a result of one of the more bizarre microscopic forms of conflict that can take place far below the power of the eye to see.

As we have developed sophisticated tools to fight, control, and frequently simply annihilate bacteria or other forms of microbial life, they often have developed their own pretty good strategies to kills off their own competition, and besides causing us occasional intestinal woe can be rather good at picking on someone their own size. Obviously, the production of fermentation products such as acid and alcohol can be a good way of clearing the neighborhood, but other substances called bacteriocins, mentioned earlier when we talked about kefir, can be produced that are essentially microbial antibiotics and can be used by one type of bacteria to kill another. The best known of these to be extracted and applied by addition to food for preservative reasons is called nisin, produced by the lactic acid bacteria we find in fermented dairy products; it was first isolated in the 1930s, and today is used in products such as processed cheese and canned foods because of its ability to suppress the growth of undesirable bacteria, particularly the spore-former *Clostridium botulinum*.

However, lactic acid bacteria sometimes encounter an attack from an unexpected source, which is a type of virus called bacteriophage, which can lead to the cheesy catastrophe previously mentioned. These bacteriophages are quite astonishing entities, frequently seen under a powerful microscope to have a structure reminiscent of an Apollo moon-lander or a hypodermic syringe. Their single-minded purpose in life is to commandeer bacterial cells by injecting (hence the syringe analogy) their genetic material, which then hijacks the cell's own replicative functions to turn said infected cell into a factory to make more bacteriophage, synthesizing the proteins that make up its outer shell and filling it with more nucleic acid. Once enough are made, the cell bursts, releasing the host of progeny viruses, each of which seeks to infect another cell, and so an infection can spread and escalate rapidly, until the culture is dead and the vat fails.

The eerie swarming of a bacterial cell by bacteriophages, clustered on its surface while they attempt their ram-raid, can be seen in Figure 7.4. As the old saying goes, it really is a case that "great fleas have little fleas upon their backs to bite 'em, and little fleas have lesser fleas, and so ad infinitum."

Interestingly, there is another type of molecule that does what bacteriophages do, which is kill bacteria, and that is antibiotics. For this reason, milk collected for cheesemaking is usually tested to make sure that it doesn't contain antibiotics (for example, if produced by cows being treated by a vet), in case these would kill or slow down the starter culture added to kick-start the fermentation.

Bacteriophages were recognized for their medical antibiotic-like properties even before actual antibiotics were available, and a century ago "phage treatment" for bacterial infections was a known medical practice. Although this fell out of common practice when chemical antibiotics became readily available, it is actually being revisited now, as bacterial strains resistant to common antibiotics have become a source of significant global medical concern. A number of reports in recent years have indicated that the extreme specificity of bacteriophages for particular types of bacteria suggests that, for bacterial illness, what goes around might come around, long after the first use of these microbial warriors.

Fermentation at home

Perhaps not surprisingly, given how long fermentation has been (even accidentally) practiced, fermentation is something that can be practiced in almost any environment without complex equipment (the main requirement being patience), particularly if that environment comes complete with convenient safe bacteria or yeasts to do the job, as mentioned already for kefir and yogurt.

Such "subculturing" methods might not even be necessary, however, for fermenting vegetables, as is quite popular at present, because the raw materials often come pre-inoculated with their own cultures adapted to the task. Where fermentation is not desirable, of course, the goal is to stop the action of such microorganisms by refrigeration, cooking, or otherwise preserving, but a whole new set of flavors and character can be generated by letting them off the leash to follow their natural fermentative urges.

The most common fermentation pathway in many vegetable fermentations is that which leads to acid. This is essentially getting to the same result as pickling in vinegar, as acetic acid can be a common fermentation product (alongside lactic acid), but allowing the pickle to grow itself, as it were.

Of course, taking a wild-type natural microbial population and positively encouraging it to grow involves some risk that no undesirables are present that might leap similarly at the chance, so in such cases salt is often added to control the less-desirable fellow travelers, possibly through pickling in brine, where the combination of salt and a (relatively) oxygen-free environment will lead to an acidic fermentation.

These are the basic principles of making sauerkraut, where finely cut cabbage is held in brine for a few weeks under slightly warmer conditions than room temperature, while leaving chunks larger and adding spices such as chili and garlic to the brine will give kimchi.

A less-complicated option might be to use a commercial starter culture to make sure that all that grows is that which you want to grow. Yet another option involves going to a source of plenty of bacteria that like to ferment sugars to lactic acid—such sources include many dairy products, such as yogurt and cheese. Simply straining such products (perhaps after mashing up the cheese in water) will result in a culture-rich fluid that can be used as a highly effective starter for vegetable fermentation.

Home-brewing and wine-making have also been popular for many years, and indeed fruit and other sugar-rich materials can be fermented to alcoholic ends, with the alcohol in this case giving the preservative effect, as it deters the growth of most undesirable undesirables. Almost any fruit can be pulped and strained to give a fermentable mash, to which yeast can be added and the mix allowed to ferment, again over a few days at a suitably warm temperature.

Wherever we look, bacteria and other microorganisms are all around us in the world, sometimes helping us, sometimes hurting us, and sometimes ignoring us, and their presence has been an unacknowledged part of the processing, preservation, and problems of food since we first consumed food. Advances in microbiology have put the science behind the observations, and now we can control this unseen world to unprecedented degrees. In later chapters we will focus on how processing of food is most commonly undertaken with the objective of killing those microbial species that might otherwise cause us ill (literally), but first we focus attention on one form of microorganism that plays a key role in many familiar food products: yeast.

8

The Rising Power of Yeast

In the last chapter, yeast was mentioned a few times as one of the generally less-problematic microbial denizens of food systems, and in fact the roles of yeast in the production of two of our most common and popular food categories, alcoholic beverages and bakery products such as bread, are so critical that it is worth dedicating a whole chapter just to the consideration of the science of these products.

Alcoholic fermentations

The ability of yeast to grow in a wide range of raw materials and convert sugars to alcohol, carbon dioxide, and other interesting products is the basis for production of products such as wine and beer, as well as higher-alcohol-level spirits, and is a process that has been exploited for the purposes of human pleasure for thousands of years.

The origins of alcoholic fermentations, like those of many food products, are somewhat murky, but it is thought that honey or fruit may have been the original basis for the fermentation of such products, and that wine arose because of accidental adventitious spoilage of grapes and their juice that turned out to have, well, interesting consequences. The Greeks and Romans had wine-making down to an art, and it features frequently in their art; it also makes many appearances in the Bible (including a nonscientifically verifiable production protocol based apparently solely on water).[1]

The main reason alcoholic fermentation became of interest was as a way to prevent bacteria or other undesirable microorganisms from growing in juice by allowing a different kind of microorganism to get there first, use up the goodies, and produce products that made conditions highly unsuitable for colonization by later invaders. We routinely associate the word "intoxicated" with a formal description of the result of overconsumption of the outputs of such fermentation, but

---

[1] An early biblical strain may have been called yeast of Eden.

the heart of that word is "toxic," which reminds us that alcohol is a poison. It just happens to be one that humans can tolerate only up to certain levels, beyond which poisoning and death can readily occur, but at lower levels has a range of effects that need not be described here. For bacteria, however, the key effect is the toxicity, and so even low levels of alcohol will prohibit their survival.

Today, we are familiar with wine as a wide-ranging term covering a vast range of brands, breeds, names, and prices, but falling in the most broad terms into two main categories, which are white and red, with rosé as a third less-common offering. The difference between white and red wine starts with the grapes, with two main colors of grapes yielding the respective wine types. Overall, there are more than 10,000 different strains of the vine plants that give rise to grapes, and wine can be made from one or a mix of the extracts of these grapes, but a fundamental starting question is whether the grape is dark (red) or pale (green).

There are a number of compositional differences among different types of grape, which are reflected directly in the final wine's character. At a most basic level, red grapes contain molecules called anthocyanins and tannins that are found largely in the skin of the grapes, and so the color of the final wine depends to a certain extent on the extraction method used to separate fermentable juice from skins.

Slight differences in extraction procedure and final level of acidity (pH) after fermentation result in the fact that red wine actually comes in many (probably more than 50) shades of red. Anthocyanins change color quite dramatically as a function of pH, going from red or pink in acidic conditions to greenish-yellow in more neutral conditions and even a colorless state in very alkaline conditions.

In terms of the fermentation and ultimate flavor of the wine, another consideration is the final balance between acidity and sweetness. The more sugar that is present initially, the more alcohol that can be produced, and when all the sugar has been converted to alcohol the wine is described as "dry"; wines in which some unfermented sugar remains are referred to as "sweet."[2] In parallel, the level of acidity affects the flavor and character of the wine, with excessive acidity giving a sour flavor, while too little can result in a "flat" flavor.

The levels of sugar and acidity present in the grapes also change as they ripen, with acid levels decreasing and sugars increasing, which means that the grapes must be harvested at exactly the right point so as to start the fermentation with the best possible set of initial conditions and give a higher likelihood of reaching the best possible end point. Red wine grapes are typically harvested later than white wine grapes, and hence have a higher sugar content.

The ideal final pH for wine should be around 3.0, which is more acidic than yogurt. Such a low pH, plus the alcohol present, provides an incredibly hostile environment for the growth of microorganisms, which is why wine can last for such long periods of time without fear of spoilage or safety concerns.

---

[2] Normal opposites such as "dry/wet" and "sweet/sour" clearly don't apply in wine terminology, and anyone asking for a wet wine might be met with understandable bemusement.

Overall, what we taste and smell and see in the glass depends to a huge extent on the internal chemistry of the grape at harvest, which in turn depends on the climatic conditions that led to that point, and how they sped up or slowed down chemical reactions within the grape. These climatic factors include average temperature during growth and extent of direct sunlight, which directly influence these reactions. Hot temperatures close to harvest will favor more sugar production and less acidity, giving strong but flat wines, while cooler temperatures might have the opposite effects. This is one reason why vintage (year of harvest) can have a notable effect, because of year-to-year differences in weather conditions at the point of harvest. In addition, this explains why different regions (with even tiny differences in climactic conditions to those of other regions) can produce different types of wine. Different soils in each region also influence the flavor profile of the wine.

So, the difference between wine types is, among many other things, due to the combined effects of specific grape grown, the soil in which it grew (in terms of content of slate or chalk, or drainage, for example), the temperature during its growth and in particular at the point of harvest, and then the conditions applied during its processing and fermentation. The exact same variety of grape, grown in different soils in a different year, can give a different wine at the end; a French Pinot Noir can taste different to a Californian one, and 2010 might be a better year for it than 2009. The elevation of growth and shelter from wind can also be key factors, as chill winds can slow down reactions at key steps, while more direct exposure to sunshine can shrivel grapes into raisins.

Once harvested, the wines are crushed to extract their juice (referred to as "must"), filtered to remove the skin material, fermented, and then aged or matured. While we may be visualize images of grapes being crushed underfoot in a traditional extraction, today this is likely to be done by mechanical systems designed to remove the stems and separate juice from skins as efficiently as possible, with processes for white wine needing to be more discriminating in terms of removing skins than those for red, in which case molecules from the skin are desirable to capture and include in the must (one could say that they are a must for must).

The agent that causes the fermentation may be wild yeast on the surface of the grapes, or, to ensure more consistency and predictability of fermentation, an added culture of carefully selected yeast. The growth of undesirable strains of (wild) yeast can be rigorously controlled by addition of a gas called sulfur dioxide, which has the added bonus effect of acting as an antioxidant, inhibiting undesirable oxidation reactions in the wine that would result, for example, in brown rather than red wine being produced. The fermentation typically proceeds for 8–10 days at temperatures slightly above room temperature (20–30 °C), this temperature being usually a little lower for white than for red wine.

The chemistry of the wine fermentation will differ from variety to variety, as previously explained, and is also influenced by the temperature at which it is carried out, with lower temperatures allowing higher alcohol levels to be reached, while deterring unwanted microbes from growing. The tannins I mentioned earlier play a key role during fermentation, as they stabilize the pigments in red wine, contribute

an astringent flavor, and importantly form complexes with (and cause to precipitate during fermentation) large molecules such as proteins and carbohydrates that might otherwise cause cloudiness in the final product. They can be thought of as a large molecular net that entraps other large bulky molecules and drags them downward, where they settle out like autumn leaves falling from trees to form a carpet on the ground, and can be decanted or removed by filtration, or even as a molecular Hoover scooping up undesirable particulates from the wine. The level of tannins is much higher in red than white wine (by a factor of around 10), and darker wines contain more tannins.

Throughout fermentation, as alcohol is produced, so is carbon dioxide, and fermentation and aging tanks may be fitted with valves to release this accumulated gas at regular intervals. The carbon dioxide also has a further effect, taking with it as it rises light solids in the juice, such as grape skin particles, which then form a layer on the surface of the liquid after fermentation, in a mirror image of the carpet of heavier settled materials just mentioned. To make sure that the right amount of pigments and other desirable components are extracted from the skin, mixing might be applied during fermentation to push these floaty bits back down. The more sparkling wines are encouraged to reach a much higher level of carbon dioxide than their less-sparkling relatives, and sugar may be added to the fermentation (or added with fresh yeast into the bottle, for a second go at fermentation) to allow this (or, in cheaper versions, the wine itself may be directly carbonated).

Other products of fermentation include acetic acid (more familiar home: vinegar), acetaldehyde (more familiar home: yogurt) and glycerol (a common food additive, which contributes significantly to the mouthfeel of wine). Some varieties, such as Chardonnay, derive their specific flavor from the presence of single molecules not found as predominantly in other types, in that case diacetyl, another compound that is associated with buttery flavors in dairy products.[3]

While the main actors in the fermentation, the stars of the show, are the yeast, with their miraculous ability to turn water (well, sugary and acidic water) into wine and generate most of the flavor of wine through their metabolic actions, bacteria may play a key supporting role. In the so-called malolactic fermentation, lactic acid bacteria (commonly found in cheese) transform acids in the juice (malic and tartaric, the former more familiar as the source of sour flavor in green apples and the latter being a key component of tartar sauce) to the softer-tasting lactic acid. This maintains the level of acidity needed in the wine while changing the specific mixture of acids that causes that acidity and making the wine less sour and "tart" (when it comes to wine, not all acids are equal!). This reaction starts when the sugars have been consumed by the yeasts and can take place over 2–4 weeks. When it is not desirable to have the malolactic fermentation, the wine is cooled more rapidly after fermentation, which may be the case for white wines for which a more tart flavor

---

[3] It is noteworthy that so many different fermentation processes for different food products end up with the same soup of molecules being produced, and great differences in flavor simply depend on balance and whether they are major or minor contributors to the overall sensory experience.

is desirable. A reported downside of the malolactic fermentation is the production of compounds called biogenic amines, which have been associated with headaches (clearly the impact is dose-dependent, in terms of consumption levels).

Following fermentation, the wine is allowed to settle (or racked), during which step yeast and large particles settle out and the clear wine removed; this might be repeated several times during the first year or two after manufacture. Wine barrels used for this step are typically made from oak, which presents an additional source of flavor molecules.

In Chapter 6, we discussed pectin as a major carbohydrate present in fruit, and this includes grapes, so pectin can be a major contributor to cloudiness in wine, and sometimes agents (enzymes) are added to break down the pectin, reducing the chance of this defect occurring. In some cases, after fermentation, the wine may be filtered to help remove cloud-promoting substances or cooled to very low temperatures to promote their precipitation and thus simplify their removal (this can include crystals that form from tartaric acid). Agents might also be added to help to clarify the wine by further promoting the precipitation of undesirable substances, and these might include fine clays or even proteins from unexpected sources such as milk and fish.

This aging stage following fermentation can be undertaken in tanks, casks, or bottles, and also involves a small amount of oxygen being allowed to react with some of the products of fermentation and prevent the development of undesirable flavors (the nature of this can differ between casks, through which air can permeate, or bottles). Louis Pasteur was the first to systematically study the fact that allowing air to react with red wine caused the formation of a brown sediment, and so clearly a little oxygen is good, but too much can be a bad thing. Some evaporation of compounds can also take place during this step.

And thus, step by step, year by year, wine moves from grape to glass, from must to must have, from vineyard to vintage. No matter what hazy mythology can be built up around all the factors involved in making wine (from terroir to climate), these all can be explained in purely scientific terms because of their influence on the chemistry inside the grape, and thus the exact complement of molecules that either directly contributes to the wine flavor or provides a precursor to be transformed during fermentation into something interesting. Scientific correlations could be developed between elevation (and hence temperature), sunshine, soil components, and even microbial ecology, and the growth on the outside of the grapes and the precise list of molecules present, and their relative abundances. The entire field (vineyard?) of viniculture can thus be collapsed into a laundry list of molecules, just as wonderful but no longer as mysterious.

Bread and baking

Bread is another product that we mightn't intuitively think of as being fermented, but that is certainly dependent on the activity of tame and friendly microorganisms for its creation, again yeast, but in this case for the purposes of producing carbon dioxide rather than alcohol.

Making bread starts with a mixture of flour, water, salt, and yeast (typically a species called *Saccharomyces cerevisiae*, the same as used to make ales and wine), which are kneaded together to form a homogeneous mass, which is then allowed to rise and thereby increase in size, before being baked into a solid final structure, during which stage the desirable flavor and color (thanks to the Maillard reaction, among other changes) are also developed. When flour and water mix, the starch and proteins present absorb water and dormant biological agents such as enzymes are awoken and become active, converting the large starch molecules into smaller simpler sugars, which feed the yeast.

During the rising step, bubbles of carbon dioxide are produced through the fermentative efforts of yeast, and the dough rises. These bubbles could easily be lost to the air if they were not somehow trapped, so they must become parts of the structure, with that structure expanding as a result into a light, airy structure inflated from within by the presence of the gas.

But what traps the carbon dioxide within the dough? This is where a protein complex called gluten comes in.

When considering the quality of bread, a traditional test for acceptability is to squeeze the outside of the package; good fresh bread should feel "springy," and compress under our touch but yet rebound, while bread that is stiffer to the touch will be seen as being stale, and thus less desirable. What is present in bread that gives it this springiness? It happens that this desirable structural characteristic is due to a protein (as we have seen already to be the case for so many food products), and in this case the protein complex responsible is called gluten.

All over the world, the incidence of coeliac disease is rising. Sufferers of this condition have an allergy to gluten and must avoid traditional baked products in which gluten is present and consume instead specially formulated products, whether bought in a shop or prepared in a restaurant, which have been produced without gluten. In such products, other molecules must have been pressed into service to serve an equivalent function.[4]

But what does gluten do? Coeliac disease is often referred to as an allergy to the protein gluten in cereals such as wheat and thus in products made from such cereals, but in fact gluten is not strictly a protein, but rather is a complex structure formed by the interaction of two different proteins, glutenin and gliadin. The former has a structure composed of long chains of amino acids, while the latter is a shorter

---

[4] This is a very good example of how a clear nutritional imperative can be stated for very sound scientific grounds, such as removing gluten from a recipe, but the scientific and technological challenges required to deliver this outcome are enormous. Leaving something out calls into question why that molecule was present in the first place, and sometimes one component can be found to be like a key supporting pillar in a structure that, if removed, leads to collapse, or at least major problems. Another example of this is the challenge of removing salt from food, or reducing its level, again for very good nutritional reasons, but complicated by the fact that salt typically serves multiple functions, from flavor to preservation and even modifying the behavior of other constituents, such as proteins.

molecule that forms linkages with its stringier partner, and forms a cross-linked structure whereby the gluten complex can be stretched into different directions and shapes. Kneading dough pulls the complex out straight, while baking coagulates the protein complex (by denaturing it) and "sets it" in place. Gluten is like a moving machine part and, like most machine parts it works more smoothly and moves with less resistance when well oiled. In this case, the lubrication can come from fat present in the bread formulation. The kneading also helps to disperse the gas produced by the yeast into smaller cells and gives a fine network of tiny bubbles of carbon dioxide and air.

The means of producing different baked products have been known and practiced for centuries and are shown in Figures 8.1–8.4. The raw materials shown in Figure 8.1 are first mixed, and a dough is formed and kneaded, as shown in Figure 8.2.

The microscopic images in Figure 8.5 help as before to visualize what is going on. In the original wheat flour in the top image, we see a range of particles of different sizes, from large boulder-like starch granules to smaller clumps of what are likely protein and fiber.

The production of bread is a great example of a very traditional food process, the practice of which in kitchens has probably not changed much for centuries. So many of the actions undertaken in a kitchen, such as kneading dough, and the different approaches to be taken when making pastries compared with bread, will be found in recipe books stretching back far before the identification of the proteins present, let alone the relationship between their molecular structure and the texture of the resulting product. Nonetheless, kneading and stretching at a macroscopic level is fundamentally an act of manipulating proteins, just lots of lots of them so that the manipulation can be done with flour-dusty hands rather than microscopic probes. In the middle image in Figure 8.5, we can just about see the appearance of fine strands of protein encircling the starch granules in a micrograph of kneaded dough.

Here it is useful to introduce one of the more unusual textural descriptions of food products, which is whether they have a "long" or "short" structure. We associate the term or prefix "short" with crumbly products, which won't withstand much physical pressure (bending, pulling, stretching) before rupture, like shortbread biscuits that are always associated with a high propensity to disintegrate into a mess of crumbs at the slightest physical insult. At the other end of this odd scale of structural length are "long" structures, which can be bent, pulled, or otherwise stressed to much greater extents and still return to something approaching their original shape when that force is removed. An elastic band is obviously the ultimate long structure, but processed cheese, or even bread that compresses but bounces back, is likewise long.

In baked products, we can achieve a long or short texture by stretching the gluten complexes into more or less ordered structures; knead the bread very little, and little structure is built up and a short texture results, giving a crumbly pastry, for example. In the case of pastry, in fact, we specifically design a structure in which thin sheets that are high in gluten are separated by layers of fat (kept solid enough to not melt into the gluten-rich layers by keeping temperatures low) and rely on the generation of steam from water in the mix when we cook to gently push apart the layers, giving fine multilayered structures.

If we assume that the structure is built of protein and imagine designing an optimal arrangement of protein molecules to behave in a certain manner, then we can imagine that other components that make their own structure and break up the protein network and prevent it from achieving its most ordered potential, will make achieving the more complex long structures much harder. This is just what happens when solid fat is present, and as a result we refer to the inclusion of fat in such products as "shortening," because it prevents long structures from forming, and that is the long and short of it. It is important to note that the fat that acts as shortening must be solid at the temperature at which kneading is taking place, because, if it is liquid, it will oil and lubricate the mass, as mentioned earlier. Solid lumps of fat, on the other hand, will just sit there being odd-shaped and obstructive, getting in the way, and presumably frustrating the hell out of the protein molecules as they try and stretch and order themselves (picture someone preparing to do stretching exercises on a mat in a gym to find their mat lumpy because bowling balls have been sewn into the lining!).

Butter is a typical shortening agent and appears in the ingredients image shown in Figure 8.1, because the fat present, as we discussed in Chapter 6, is normally at least partly solid at room temperature. Vegetable fats that have been made harder by hydrogenation (adding hydrogen atoms to transform them from soft unsaturated fats to harder saturated ones by increasing the temperature at which they melt) can also be used as shortening agents, but hydrogenated fats are less recommended today because of health concerns about *trans* fats they may contain.

Going back to the bread manufacture, one obvious question is where does the alcohol go and why there aren't limits to how much bread one can eat before driving or other activities are forbidden. While in both alcoholic (like beer or wine) and bread fermentations, the key reaction is yeast acting on sugar, the key difference is the presence of air that, being more available during a bread fermentation than when a liquid is being fermented, drives the reactions the yeast catalyzes through slightly different pathways. These lead to more carbon dioxide being produced than alcohol (and the little alcohol produced is evaporated during baking); this is why we don't end up with something that could more imaginatively be called beeread.

These reactions take place after the kneading is complete, when the dough is shaped and then set aside for an hour or so in a warm place to allow the fermentation to proceed (this step is often called proofing). During this time, the yeast does its business, converting sugars to carbon dioxide and minor flavor components, and the progress can be seen clearly by the expansion of the dough, to give the expanded floury shape shown in Figure 8.3.

If the fermentation also includes the growth of lactic acid bacteria, sourdough will result, giving an extra tangy taste because of the production of lactic and acetic acids, and a resulting drop to a pH value that can be lower than that of yogurt (giving sourdough bread a longer shelf life than that of other types). In traditional sourdough production, the cultures of yeast and bacteria responsible for the fermentation are "wild" and come from the bakery environment or raw materials themselves, where they have become well established over long time periods, and so flavor and character can be highly individual to different locations and their unique starters. The wild yeast found in sourdough is more likely to be *Saccharomyces exiguous* or *Candida humilus* than the *Saccharomyces cerevisiae* used in other bread fermentations, while one common sourdough bacterium found first in a certain American city (guess which one) is called *Lactobacillus sanfranciscensis*.[5]

In other bread variants, yeast isn't used at all as the source of carbon dioxide, and instead sodium bicarbonate and acidified buttermilk (or sometimes, in traditional recipes, yogurt) are mixed with flour and salt. The lactic acid in the buttermilk reacts with the sodium bicarbonate to produce carbon dioxide, and thereby produces the gas bubbles that cause the dough to rise (while more carbon dioxide is produced during baking, as heat breaks down more of the baking soda). Replacing biology with chemistry in this way speeds up the bread-making process and makes it more foolproof, allowing the use of simple and cheap ingredients. While chemical-driven alternative food processes might be regarded as a recent innovation, so-called soda bread made in this way has been made for centuries in many countries, and remains a very popular product in Ireland, in some recipes being made with a stout like Guinness as an ingredient to add color and bitter character.

Another key component of any bread formulation is salt, which, as in many food products, serves multiple functions. Yes, it contributes (mildly) to flavor, but its main purpose is actually as a modifier of both microbiology and chemistry in the product. It can modify the structures that the gluten produces (by introducing a lot of electrical charges into the mix, literally, and binding to oppositely charged regions of the proteins, thereby affecting the interactions and structures those molecules can participate in). In addition, it is a potent inhibitor of microbial activity, so can prevent undesirable bacteria from growing in the mix and also limit the activity of the yeast (for which reason it might be added a little after the yeast, not to dampen down its bubble-producing efforts before the dough has had a chance to rise).

Of course, when we think of the structure of bread, and in particular how it changes on storage, we cannot overlook the role of starch. The next stage after fermentation is baking, when the risen loaf is placed in a hot oven for around half an hour. During the early stages of baking, the yeast is killed by the heat, and then a series of complex chemical reactions take place, including gelatinization of the starch, denaturation, and coagulation of the protein matrix, and reactions between sugars and proteins (the Maillard reaction, often encouraged by including a source of milk solids in the recipe) that develop the characteristic dark brown color of the

---

[5] The names of fermenting microorganisms in food can be quite literally related to their origin. A certain beer-fermenting yeast was at one time called *Saccharomyces carlsbergensis* (probably the best yeast name in the world).

crust (as seen in Figure 8.4). In addition, the gas bubbles expand in the heat, and the bread rises further, just as the bubbles are set in their size and shape by solidification of the starch–protein matrix that surrounds them.

In the bottom image of Figure 8.5, we can see a far less defined and more chaotic microstructure, in which the starch granules have become less defined, the material has shrunk because of the loss of moisture during baking, and the mass has become dominated by a matted matrix of coagulated protein and fiber swollen by the absorption of water.

The key role of starch in the process is evident from its prominent appearance in the material at all stages when studied under our electron microscope (Figure 8.5). As we cool the bread after baking, the starch starts to crystallize immediately and, as the bread is stored, this process continues slowly (where it is called retrogradation), until at some point the crystals have interlocked such that they have essentially overwhelmed the protein-based structural elements, and the bread appears unacceptably solidified and has arrived in a state that we call stale. When we baked the bread, we destroyed the crystalline structure of amylose and amylopectin, but given enough time they want to reestablish some form of structure, and so rearrange and align their molecules to do so, forming new bonds and sometimes exuding water in the process. The more structure such crystals have formed, the more stale and solid the bread will appear and the greater the loss of its gluten-derived springy structure.

This is reversible by one simple step, however, that remelts the starch crystals and reverses the retrogradation process—this is toasting, but is accompanied by other reactions, such as the Maillard reaction and further dehydration.

In conclusion, bread is another lovely example of a product for which art long preceded science, and now we appreciate the science involved as being thanks in part to the biological properties of living yeast on the one hand, and the chemical properties of components such as proteins, fats, and salt, as well as the changes that we induce when we subject the whole complex mix to high temperatures, whether in plant or kitchen, on the other hand.

9

A Word on the Wonderful Weirdness of Water

Before we move forward from our previous chapters' exploration of the importance of the microbiology of food from its many different angles to start to focus on how we process food to, among other effects, control that microbiology, we need to consider one more basic constituent of food.

This is because, even after several earlier chapters in which the key functions of proteins, sugars, lipids, and other rather high-profile food constituents were discussed, we have yet to discuss explicitly the one that is perhaps the most significant of all. It was mentioned many times of course, lurking in the background like a supporting character actor in a movie who doesn't dominate the foreground activity but is a key part of the scene.

This magically powerful ingredient is water, yes water, that represents the majority of most food products, and without which most of their properties and characteristics would not exist. We have seen already how water can appear in food in many guises, depending on whether it deigns to interact with the other constituents present, leading to apparent logical surprises like the fact that a melon (a solid?) has actually more water per gram of its weight than milk (a liquid?), just because in one case the water is absorbed and robbed of its innate fluidity, while in the other no such restrictions apply.

Besides influencing texture in a completely fundamental way, though, water influences behavior of just about every other molecule in food, from the structure of a protein (and hence the texture we perceive) to the suspension of oil droplets in the many food products that are emulsions.

As well as this, almost all the dynamic changes we encounter in food, for better or for worse, depend on water. Microbes require water to live, as we can see when we preserve food by removing it (in drying), or else denying it more subtly by adding substances such as sugar or salt, which can suck the very water out of bacterial cells like molecular vampires. In addition, textural changes from the staling of bread to the softening of forgotten biscuits depend on the behavior and movement of water. Likewise, changes from browning of fruit to softening of cheese curd into desirable textures during ripening are driven by a particularly active type of protein called enzymes, which again require the presence of water to act.

This critical role of water has led to a number of terms and parameters being used to reflect not just its level but its actual behavior in food. At the most basic level, we can characterize food based on its water or moisture content, crudely determined by heating a weighed sample food in a suitable oven or chamber (for example, using a conventional or microwave oven) to remove all the water, then reweighing and calculating the level of water present by difference. This allows us to work out the absolute level of water (and, by subtraction, solids), but doesn't tell us much as to what the water was actually doing in that food.

To understand such actual behavior, food scientists refer to what is called the water activity, which, as the name suggests, measures how active the water is (or alternatively we could think of it in terms of how much other activity that water can permit or encourage). Two samples of food could have the same level of water but very different water activities and different stabilities. Many food products (including milk, meat, fruits, and many others) have a water activity close to 1, which means essentially that 100% of the water is free to do its own thing but, when we dry food, or reduce it by the means mentioned earlier, the water activity drops. Honey might have a water activity of around 0.5, while a biscuit might be up around 0.6–0.7 and a dry coffee down around 0.2–0.3.

The key point is that *most* undesirable reactions of consequence in the food are less favored as the water activity falls, and so reducing this activity can be key to preserving the food. This explains why drying or sugar addition, to take but two examples, have been such effective preservative tactics for so long. When I said *most* in the previous sentence, the key exception is oxidation of lipids, which actually is favored at lower water activities (perhaps because the molecules on which the oxygen must react to induce rancidity or other problems are essentially held in place in the absence of the mobilizing powers of water, and the oxygen doesn't need water to move around and wreak its nefarious effects).

So, from the perspective of understanding the characteristics of food components and ultimately how these build up to give the properties of actual food products, as well as how the quality of that food changes over time and is biologically dynamic, water is the key.

The focus for the rest of this chapter, though, is less on the biological than on the physical properties of water, as understanding these is critical for understanding how we process food to make it safer and more stable, which will be the subject of a number of later chapters.

As we will see, water is actually a pretty weird material, which behaves in many ways that are different from those of any other substance, and understanding this is key to understanding food preservation.

The changeable states of water

To start with, we encounter water in three different forms in the world around us (of which food is of course the part we are most interested in): solid ice, liquid water, and gaseous water vapor or steam.

In each case, the water molecule is composed of one atom of oxygen bonded to two atoms of hydrogen, giving the molecular formula known to the vast majority of people who don't know a single other molecular formula, which is H_2O—two hydrogen atoms, plus one of oxygen.[1]

To understand the properties of this molecule, we need to picture its structure. The hydrogen atoms are bonded to the central oxygen in a way that could be visualized as sticking out like a rabbit's ears. We don't need to get bogged down too much in the intricacies of atomic bonding, but need to just know that the bonds between the atoms are formed by sharing components of the atoms called electrons, which have a tiny electrical charge. Electrical charge is one of the key properties of atoms and molecules, and for our purposes let's just say that electrons have a negative charge, while the other main charged components of atoms, protons, have a positive charge.

The bonding arrangement in water means that electrons are shared between the hydrogen and oxygen atoms in an uneven way, and the molecule has a difference in charge at different parts of its structure. The electrons are basically being tugged away from the hydrogen atoms toward the oxygen atom. A hydrogen atom is the simplest of atoms and includes just one proton and one electron, but, when bound into a water molecule, that single electron is shared with the oxygen atom in a chemical bond but is held a little more closely to the oxygen atom than its parent hydrogen. So, the oxygen part of the water molecule has a slightly stronger negative charge (thanks to the two electrons it is sucking in from its two partner hydrogens), while the hydrogen atom ears of the molecule are more dominated by the charge of the remaining proton and are slightly more positively charged.

A key principle in physics is that similar electrical charges push each other away, while opposites attract.[2] So, when water molecules come close, the hydrogen atoms on one molecule can be attracted to the oxygen atoms on another, and so weak attractive bonds between the molecules can readily form. In addition, the water molecules are pretty small,[3] and so large numbers of them can fit pretty neatly together in different shapes stabilized by these attractive forces.

These atomic electron squabbles lead to many of the key properties of water, as they influence how much individual water molecules attract each other and make them cling to each other, which is entirely dependent on how much energy is present in the system. Energy makes the molecules move around and can either encourage or discourage their bonds of mutual attraction.

---

[1] A mischievous teenager once organized in a campaign to ban an extremely dangerous-sounding compound called dihydrogen monoxide that, among many crimes, was found in acid rain and chemical weapons, caused burns in both its solid and gaseous forms, caused soil erosion and death if inhaled even in small quantities, and yet was found in or added to many common food products. Clearly, based on this description, one can only say "down with that sort of thing," and luckily the ironic subtext was recognized before the banning of water gained political support.

[2] Think of this as how the similar poles of two magnets can't be pushed together because an invisible force pushes them apart, while the opposite poles can readily be held together.

[3] Obviously, the water molecule is actually mind-blowingly small, but here the term is applied relative to just about any other molecule found in food and not, for example, relative to an elephant.

At very low temperatures (below 0 °C), there is very little energy in the system, and water molecules move very sluggishly and so have time to interact with each other so well that they link up (thanks to the electrical effects previously described) into a highly ordered lattice-like structure, in which they essentially stay exactly in place, tied to their neighbors; we call this ice. This could be envisaged as a crowd of people standing very close to each other by wrapping an arm around each neighbor, holding them close, and making large-scale movement beyond a glacial shuffle rather difficult.

When we warm ice, we add energy to the system and jiggle the molecules more and more. Eventually, we weaken the power of mutual attraction, the lattice breaks down, and the highly ordered structure disappears. The ice melts and becomes liquid water.

However, the water molecules have still enough electrical attraction for each other that they want to stay nearby and sense strength in numbers. Where one goes, the others are likely to follow, but with good individual mobility, so the water can slosh and pour and flow, but remains water. Taking the previous analogy a step further, the arms around each other have been replaced by held hands, and the crowd is much more mobile, while retaining the same number of people staying together.

Add more energy, though, and each molecule/person moves more and more rapidly, and the hand-holding stretches, as the bonds between water molecules are stressed and stretched. Eventually they rupture, and the individual peoplecules are no longer part of a single mass, but free to go off on their own, flying off in their newly highly energized state wherever they wish.

Water has become steam (if water boils) or water vapor (if it has evaporated at lower temperatures).

The transitions between each of these states has required a particular threshold level of energy to be passed, and energy beyond that required just to increase the temperature has been consumed to fuel this change between states or phases.

In addition, if we run the film backward by cooling steam or water vapor, we gradually coax stem molecules to slow down enough that they get sufficiently close to their neighbors to first hold hands (so the steam condenses to liquid again), and then enough that they embrace with arms (as that water freezes), and at each stage the system must jettison bursts of energy to allow this to happen.

A quick word on surface tension

The affinity of water molecules in the liquid state for each other, above all else, gives rise to a critical property of water called surface tension. Surface tension is the reason that falling water forms a spherical droplet, because the affinity or cohesive force of the water molecules at the surface results in an equal inward pull of every part of the surface, and the surface is minimized as a result. The shape with the smallest possible volume for a certain volume of water happens to be a sphere or droplet.

Surface tension also explains why such water droplets hitting a solid surface tend to remain in droplets rather than dispersing into an even layer. At the interface of water in a raindrop with a window, for example, the attractive forces between water molecules at the surface of the droplet all point inward because of a complete lack of affinity for either the gas molecules in the air around or the molecules in the glass. The resultant surface tension can be strong enough to allow a light insect or object to float on the surface of water rather than sinking if the weight is not enough to overcome this tension.

In a food context, surface tension (or the related property of interfacial tension, which describes attractive and repulsive forces acting at the interfaces between water and phases like gas or oil) is a hugely important consideration in how powders react when mixed with water, for example, and many food constituents or ingredients we have already met exert their effects through modification of surface tension. This is a key aspect of the roles of amphiphilic molecules (those that have both water-repelling and water-attracting parts in their structure) in emulsions and foams, for example.

The manipulation of surface tension is in fact vital to the behavior of one complex chemical found in significant quantities in every food plant or kitchen, but that is not considered otherwise in this book, which is the detergents we use to clean up afterward. The key ingredients in detergents like washing-up liquid are amphiphilic compounds called surfactants that coat otherwise water-repelling food residues, particularly those that are high in hydrophobic fat, with one part of their structure, while the other attracts water to the site. This action allows these "sticky" surfactant molecules, in the turbulent conditions created during washing up, to "grab" the food or dirt and remove it from the material to be cleaned, then coating it as an emulsified droplet, which cannot reattach to the dish or plate it came from, allowing it to be washed away.

The allocation of washing-up responsibilities can be a source of tension in many households, but now we can recognize that tension in this case is a good thing, once it takes place at the right surface, and not at the dinner table!

How does this link to heating food?

To heat food, whether cooking a steak or pasteurizing fruit juice, we need to exploit some basic physics principles, particularly the second law of thermodynamics, which states, in simplified form, that heat will always flow from a hot region to a cold one, as nature tries to make both have a similar temperature. This principle can be seen everywhere, as the heat exuded by hot objects, from radiators to coffee cups, flows away into the cooler surrounding air, and hot food grows rapidly cold (while cold food in a warm room conversely warms up, because of the inexorable arrow of thermodynamics pushing heat in the opposite direction).

So, to heat food, we need a source of this heat energy and to encourage the flow (or transfer) of heat between hot and cold to take place as quickly as possible. Why is speed important? Generally, when heating food there are zones of temperature that we want to avoid, and so we want to get through these zones as quickly as possible and linger in them as briefly as possible. For example, when heating something from refrigeration temperature to a temperature hot enough to kill bacteria, we pass through the region around 37 °C where bacteria most like to grow and be active. This is not a good neighborhood to hang around in, so we want to drive through it as quickly as possible. Likewise, when we get to very high temperatures, we know that the heat will have significant effects on our food-quality characteristics, like flavor and nutritional quality, so we also want to hang around here as little as possible, but for different reasons to the previous example.

So, how do we encourage heat to move into (and/or out of) food as quickly as possible? To understand this, without getting bogged down in lots of equations, some basic principles can be explained.

Let's consider a simple system where we are heating cold milk using a hot plate as our heat source. One simple example might involve a container of milk on a stove hot plate. Such a system is shown in Figure 9.1.

The first point is that the key driver for the transfer of the heat is the difference in temperature between the hot and cold materials. The bigger this is, the faster heat will flow. A Tanzanian student called Erasto Mpemba discovered in 1963 that hot water freezes faster than cold water, because of exactly this effect. While his findings were initially disbelieved, the law describing this is now called the Mpemba effect.

There is a temperature difference in our example between the hot plate and the milk and pot, and this is the irresistible driver for heat to move in the direction that will eventually, if left long enough, lead to both having the same temperature. So, heat flows from the plate to the milk, but there is a barrier between these, which is the wall of the sauce pan, which is perhaps made of aluminum. Aluminum is a solid metal, and heat flows through solids by a process called conduction, traveling across the solids in straight lines, because the material doesn't move in any particular way to accommodate the flow of heat (other than perhaps expanding a little).

Some materials are more accommodating to heat flow (these we call conductors, the most commonly encountered example of which is metal), while others are more obstructive (we call these insulators, and they include wood, ceramics, and materials containing a lot of air, like the Styrofoam in coffee cups that keep our hands from being burnt).

So, thanks to the conducting properties of the aluminum in the pot, the heat passes through the metal wall separating our heat source and milk, and first reaches the milk in contact with the inner wall (step A in Figure 9.1). Milk is mostly water, and the part of the milk we focus on for now is that water that, when it feels the heat from its nearby wall, warms. As this warms, it expands slightly, as the water molecules absorb energy and begin to vibrate more excitedly, in effect taking up more physical space. On expanding, the water becomes less dense than the cooler surrounding water, and less-dense materials rise above more-dense ones (just like we saw happens to a mixture of water and oil in Chapter 6). In this way, a little vertical current is established in the sauce pan, as the heated milk near the wall makes for the surface, where there is more room to accommodate its heat-induced excitations (shown as the curly arrows in part B of Figure 9.1).

This movement creates a current that pushes cooler liquid out of the way (rudely, one imagines, in molecular terms), downward, and back toward the wall spaces that have been vacated. This newly arrived milk then duly gets heated, and rises, and so forth, until more and more of the bulk of the liquid has been heated, and the average temperature of the milk progressively rises.

One complication caused by the presence of milk in this example is that it might burn at the inner sauce pan wall if that is hot enough, which would give a burnt layer of gunk. This could insulate the wall and slow down the transfer of heat into the remainder of the milk (and also give a burnt flavor to the hot milk). So, a key thing to do would be to help speed up the movement of heat within the milk by mixing it as heated to help transfer hot milk from the wall and bring fresh cool milk to replace it more quickly, while making heat spread through the milk more evenly in general. Such so-called forced convection is normally preferable to allowing natural circulation to do all the heavy lifting of moving heat around the system, just as (following exactly the same principles) having a fan in a warm room helps cool down air much more than waiting for natural air currents to do the job. This is because exactly the same principles of natural and forced convection govern heat transfer in gases, whether in a food context or in a warm room.

Another situation in which heat transfer by natural convection will be slower than might be desirable, and where a little helping hand is always welcome, is when the material being heated is very viscous and so hard to mix naturally, and through which heat currents will flow sluggishly, like a swimmer trying the backstroke in a pool of honey. Soups, sauces, and stews will all need to be stirred while heating to make sure the heat is evenly distributed, giving much faster heating while avoiding burning. And, of course, when we consider the stirring itself, we will instinctively reach in the kitchen for the wooden rather than the metal spoon, as the insulating properties of the former will protect our hands by discouraging the conduction of heat from hot pot to cooler hand.

So, heat flows in this system by two means, conduction through solids and convection through liquids (and gases), but a third means of heat transfer is also at play, which can be detected by a hand placed a few inches away from the sauce pan. This hand can feel the heat literally being radiated away from the hot metal (shown by the arrows at D in Figure 11.1); radiation is the third mechanism of heat transfer, from hot materials into a gas, just as a hot corrugated metal unit on a wall pumps heat into a room, which then rises by convection through the air, to give a small but measurable temperature difference between the ceiling and floor of any room.[4]

---

[4] In this system, there is a third influence on the heating, which is the evaporation of water molecules from the surface of the heated milk as the temperature increases, which actually cools the system by removing energy; we will come to this shortly.

To consider for a moment the question of insulators, I said earlier that air (for example in Styrofoam coffee cups) can act as an insulator, and indeed it can. This fact is exploited in a famous traditional dessert product common in some countries, called a Baked Alaska. In this party trick of a confection, ice cream is surmounted by a layer of meringue, which is made from eggs that have been whipped into a structure that traps large quantities of air. When such a cake, containing other layers of delightfulness above the meringue, is placed in an oven, the air layer does such a good insulating job that the heat cannot penetrate through it, and the dessert achieves the rather good trick of having both hot and icy cold regions in relatively close proximity. Thermodynamics in delicious action!

Making water disappear

During the processing of food, whether in a kitchen or in a pasteurizer in a milk processing plant, we exploit changes of state (casually flipping between ice, liquid, and vapor, the united states of water) routinely. One of the most basic operations in a kitchen, indeed, is all about changes in state, and that is boiling a kettle.

In this case, we are supplying energy (electrical) to induce a change in state from a liquid to a vapor (steam), a process that requires a large input of energy to provide the fuel to carry those water molecules into flight (this energy is called the latent heat of evaporation).

Of course, water molecules can gain sufficient momentum to escape into the air at temperatures below the boiling point, as we see evaporation leading to a cloud of water vapor at temperatures from a bathroom to a simmering pot. However, at the boiling point of water *every* water molecule suddenly gets an irresistible urge to escape, in a highly vigorous manner. The difference between evaporation and boiling is that the former occurs slowly at the surface of a body of liquid as individual molecules get sufficiently energetic to escape, while the latter happens rapidly throughout the whole bulk of the liquid, leading to bubbles of water vapor rapidly and spontaneously forming.

On the other hand, when we cool water vapor, the opposite occurs and water droplets are produced, but in this case (to balance those strict energy books) energy must be released, and the recipient of that water vapor gets a heat bonus (called the latent heat of condensation).

These phase changes are hugely important when we think about removing water from food, which we want to achieve when we dry food, from milk to coffee.

The most important consideration in this regard from the viewpoint of food quality is to dry as gently as possible, to give a product that hasn't been burnt by the high temperatures encountered and that is also easy to reconstitute when added to water. To understand how to do this, we need to think some more about what happens when water evaporates.

To think about evaporation, consider the example previously mentioned of a pot of milk being heated to boiling point; evaporation is the process by which molecules of that water escape from the surface of the liquid into the air above, in a billowing cloud of vapor, if there is no lid or other barrier in the way (escape route C in Figure 9.1).

What factors affect the rate of this escape?

One is the pressure, or resistance to escape. This pressure can be visualized as the weight of air pushing down on the surface of the liquid, presenting a barrier to that victorious leap to freedom. The higher the pressure, the more air presses down on the surface, the harder it is to escape, and the higher the temperature at which water boils, because more energy is needed to beat the opposing force; conversely, reduce the pressure and we make the leap easier, and reduce the boiling point.

These principles are seen in practice every day, as low air pressures atop mountains allow water to boil more easily, while descending below sea level (in either case presumably with a kettle and a very long extension lead) makes it harder to boil water. Similarly, in processing food, when we want to reach very high temperatures without the food boiling we increase the pressure while, when we want to evaporate more effectively while doing less thermal damage to food quality, we use a vacuum to reduce the pressure.

The next factor is the area of surface of liquid from which the water can evaporate; the greater this area, the quicker the removal will be. If we place our liquid to be dried in a thin layer on the surface of a tray, there will be a much bigger area than if the same amount of liquid were in a narrow pot in which it formed a deep column of water, and so the faster will be the evaporation. This is in part because of the greater area from which water molecules can escape, but it is also because of the greater area of contact between the liquid and the source of heat that is fueling that escape.

For any given volume of water being dried by, say, hot air, there is one unbeatable way to maximize surface area and contact with that air, much more even than we could achieve in a thin layer, which is to covert the liquid into a spray of droplets of tiny size. Each droplet can then be surrounded by hot air, and we get maximum surface area for the volume concerned, with the total surface area being inversely proportional to droplet size. So, if we can convert a liquid into a fine spray we can present the optimal format for efficient drying.

What about the air itself? It should be dry (the more humid the air is, the less air it can absorb), rapidly moving (to sweep away the moisture molecules once they escape and bring fresh dry air to the droplet surface), and hot (because the greater the temperature difference between air and liquid, the more powerful the engine driving evaporation).

All of these concepts are actually intuitive if we simply consider wet clothes being dried on an outdoor line. The best drying conditions will be dry, warm air with a good breeze to keep the clothes moving—humid, still, cool air does not good drying promote. The same principles apply in food-drying systems.

One complication comes from the unusual properties of water discussed earlier. To transform from a liquid to a vapor state requires a significant injection of energy, above and beyond that which is needed to warm up the water. This is called the latent heat and must come from somewhere, which means that, as the water droplets dry, they cool continuously as energy is sucked from them to provide this latent heat for evaporation, and we need to factor this in to the process.

This also has an everyday application we are all familiar with, which is when we leave the sea or a hot shower into a warm beach or bathroom, and suddenly shiver with cold. This is because the water on your skin is evaporating, and to pay for its escape ticket is robbing your skin of the energy it needs to pay the latent heat bill.

Exploiting the excitement of water

The properties of water discussed in the previous section are also critical to many versions of a process called ultra-high-temperature treatment of milk, which involves heating milk to around 135–140 °C for 3–4 seconds. This gives a product that, on the positive side, doesn't need to be kept in a fridge and will last for up to a year, because such extreme heating has killed all the bacteria within. On the negative side, the flavor is much more cooked than would be found for pasteurized milk, and in countries such as Ireland it is found only in places like planes, trains, and hotels (and maybe automobiles), where it is to be added to coffee, so its flavor isn't a primary consideration.[5]

To heat milk to temperatures far beyond its boiling point without, well, boiling, requires some steps to be taken to increase the pressure, because of the effect of pressure on boiling described earlier. Then, to cause least damage to the product at the very high temperatures involved requires that the milk stay the shortest time possible in the "damage zone" within which every fraction of a second lingered causes additional nutritional and flavor damage.

How can this be achieved? One method is what is called direct heating, whereby steam is injected directly and mechanically into the milk, or else the milk is sprayed into a cloud of steam (which is called infusion). When ultra-hot steam contacts cooled milk, just like steam in a bathroom hitting a cool mirror, some

---

[5] The difference between pasteurized and UHT milk is a great illustration of the power of the values held by consumers. The former product is not sterile, is guaranteed safe only when kept in the fridge, and goes off (spoils thanks to bacterial activity) in around 2 weeks, while the latter doesn't need a fridge and stays stable for up to a year. On paper, the convenience of the latter should have made the former extinct a long time ago as a far less practical means of preserving milk, but the value consumers place on flavor and nutritional value, both seen to be unacceptably damaged in UHT milk, have kept it in the minority share of the milk market in countries that have either relatively low ambient temperatures or a strong culture of drinking a glass of milk, or both (like Ireland).

of it condenses into water. To achieve this conversion, as previously discussed, it needs to jettison its excess energy (like a plane making an emergency landing jettisoning fuel and rapidly cutting its speed), and it thus releases a sharp pulse of energy into the milk within which it has condensed, giving an astonishingly rapid increase in temperature. If the amount of steam added is carefully controlled, the target temperature can be achieved through an increase of 40–60 °C in fractions of a second.

Now the milk is hot (good!) but, critically, it has also absorbed some condensed steam (i.e., water) and so is diluted (not good!). So, the processor must (after holding the milk for the requisite few seconds) both cool the milk and remove the added water. This can be done simultaneously by exploiting the latent heats again and purging the excess heat by evaporating off the exact amount of water that was just added. This evaporation is induced to occur by creating conditions more conducive to boiling, which simply involves reducing the pressure (which makes evaporation easier, as we saw earlier). So, the pressure is decreased (often by simply allowing the milk to suddenly expand into a much larger volume), the boiling point drops, water evaporates, latent heat is withdrawn from the stingy bank of energy, and the milk cools, again in a tiny fraction of a second. The milk has now gone to the UHT temperature and back in little more than the few seconds for which it was held at that temperature, and damage, while always inevitable at such temperatures, was at least minimized.

At the other end of temperature scale, the same principles apply; when water goes from a less energetic state (ice) to a more energetic one (water), heat must be consumed (the latent heat of fusion), while freezing releases energy (the latent heat of crystallization), as we will discuss in the next section.

The latent heats also play a role in other cooking scenarios, even something as simple as toasting bread. In this case, initially evaporation of water when the bread is heated cools the product, as the latent heat is removed, but once the surface is essentially dry the cooling effect is gone and the temperature starts to increase much more rapidly; reactions such as the Maillard reaction really kick in, which is why bread can seem to toast slowly, until it gets slightly brown, and suddenly can become unacceptably burnt! In another application of this principle, when we stir-fry food in the presence or absence of liquid, the presence of liquid actually reduces the temperatures during cooking, as evaporative cooling draws off energy, and so food will cook faster when dry (once kept moving fast enough to not burn and stick to the pan).

freezing

Many materials in nature and in food experience a transition between states in which they are liquid and those in which they are solid, as we have seen. When we cool melted oils or fats, for example, they will eventually cease being liquids and harden into a solid or semisolid material. As discussed in Chapter 5, different fats make this transition at different temperatures, which explains the difference between sunflower oil and butter in terms of their basic hardness under our normal conditions in a kitchen, and why one is found in a bottle and the other is a block.

Of course, water similarly makes a transition when cooled from liquid to solid, and, like fats, the solid form takes the form of crystals, which we know as ice. We see this transition everywhere from its desirable presence in a drink on a warm day to its undesirable slipperiness on a winter's morning when we are in a rush. However, the behavior of water when it freezes into ice is rather unusual and mightn't be immediately obvious, despite our everyday familiarity with the stuff.

For a start, water expands when it turns into ice, increasing its volume. Ice cubes are larger than the amount of water placed in the tray, and a bottle of beer put in a freezer to cool on a hot day and then forgotten about can shatter dangerously because of this expansion. For water to expand on freezing means that the lattice crystalline structure into which the water molecules have become locked as ice takes up more space than the original liquid form, and so a given weight of water has increased in volume; to put this another way, it has become less dense. Ice cubes float! If this did not happen, the ship *Titanic* would have made it all the way to New York, but the earth's climate would overall be a little less hospitable.

Another cunning trick water plays when it freezes relates again to its latent heat, as previously discussed. For water to transition from a more energetic state, water, to a less energetic one, ice, requires the jettisoning of a significant amount of energy, or heat. To see this in action, consider an ice cube in a drink; does it cool the drink because it is colder than the surrounding liquid and heat flows as always from hot to cold? In part yes, but an ice cube will cool the drink much more than, say, a piece of plastic of exactly the same dimensions and temperature because, as the drink warms, the ice cube melts, which consumes energy (the latent heat), and this adds extra oomph to its cooling power.

Likewise, when we consume something frozen, it melts on our tongues, and must absorb the necessary energy to do so. So ice cream cools our mouths more effectively than its temperature alone would predict (whether latent heat can fully explain brain freeze, however, remains as yet unexplained by science). It doesn't even have to be ice that melts either, as any crystalline material will do the same. In butter from the fridge, for example, most of the fat is in crystals, and when we eat it they melt (as our mouths are at a temperature above the temperature at which they melt) and absorb heat from our mouths, and so the butter causes cooling on consumption.

The next really important property of water when it freezes is the expansion mentioned earlier, combined with the fact that ice crystals are rather jagged and sharp. This makes freezing unquestionably one of the most potentially damaging things we can do to food.

Imagine a soft-tissue-like material, like meat or fish, filled with a lot of liquid water in the cells of the material and the spaces in between. Put this at freezing temperatures, and the water will gradually expand and, instead of sloshing around fluidly but harmlessly in its microscopic homes, lock into place and become decidedly spiky, with sharp pointy edges all over the place. Suddenly we get punctures, tears, and rips through our soft food tissue, as holes are poked by the expanding ice crystals, using their crystalline elbows to push for more room.

This starts to tear the material apart from the inside out, but the frozen water remains locked reasonably firmly in (a larger amount of) space by the icy superstructure that has formed within it. Remove the food from the freezer, though, and allow it to thaw, and the ice will shrink and retreat as it melts. The damage has been done though and the texture of the food irrevocably altered as a consequence. Typically, clear evidence of this can be seen clearly alongside the food, as puddles or drops of water appear because they have forced their way out through tunnels and rips that should not have been there and have escaped from inside the food to its outside.

On the one hand, for this reason, if done badly, freezing is a very damaging thing to do to food; yet, on the other hand, we know that such low temperatures are also exceptionally hostile to microbial life, and biology in general, and so are good for preserving food for long periods. So, we know the advantages to freezing food, but the key question is, how can the freezing-induced damage be minimized?

The key to success in attaining this goal lies in the fact that the size of the ice crystals that form depends on the rate at which the transition to frozen state took place. Faster freezing gives smaller ice crystals, and so freezing really quickly will convert a certain mass of water into a much larger number of smaller crystals than slow freezing, which will in contrast give a smaller number of much larger crystals. The smaller crystals are less jagged, sharp, and damaging, and will form more uniformly throughout the food, and so damage will be greatly reduced.

So, how do we freeze quickly enough to get these desirable small crystals to form? The key factor here is the rule that the rate at which the water freezes depends on the temperature difference between the food and the medium that is freezing it (remember the heat always flows from a hotter region to a colder one, so if we freeze something we are essentially sucking the heat out of it with an even colder "heat sink"). Just as in heating (only backward), the difference in temperature is the pump that causes heat to flow; the bigger this difference, the bigger the pump, and the faster the heat will flow.

To take an example, imagine a piece of meat we want to freeze really quickly. The key will be to expose it to an extremely cold medium (typically air, and the colder the air the better) and to maximize the contact between all parts of the meat and the cold air, so as to maximize the surface area over which heat can escape. The surface is where air and food meet and is the gate through which heat can move from one to the other: the wider this gate, the faster the escape, like allowing a larger crowd to move more quickly through a door or gate by opening it more widely. So, the meat will be cut to maximize its area and hung to expose much more surface than would happen if it were, for example, lying flat on a tray that blocks the lying surface from air contact.

Now, if it hangs in still frigid air, heat will be drawn out, but a warm, moist, air layer will hang around the surface and eventually insulate the meat from further heat loss, and so that warm layer needs to be removed and replaced by fresh cold air. For this reason, the air cannot be left still, but should be moving rapidly, being blown over and past the meat, to be eventually recovered, rechilled, and recirculated. Such a system is called a blast cooler.

This may sound familiar, as it is a perfect analogy of what was described earlier for drying food.

In that case, just as we considered the drying of a spherical droplet, let's consider freezing a spherical food particle, like a pea. How do we freeze a pea, and what does it mean when a package or ad claims that a certain type of pea is "individually quick frozen"? In this case, we can hardly hang individual peas one by one like the meat, but we still want to maximize the exposure of their surface to cold air.

To see how this could work, imagine a layer of peas lying on a metal tray or bed, with cold air blowing over them from above; the upper layer cools well, but their undersides remain unchilled. Now reverse the picture, and blow air from below. This could be achieved by perforating the metal bed with tiny holes, as long as the holes are smaller than the peas (so they don't fall through). At a certain (very low) air flow velocity, the peas will remain undisturbed, but turn up the air flow and they will start to jitter, and eventually they will rise magically into the air and float, in proud vegetal defiance of gravity. Turn the air flow up a little faster yet, and they will now move, flowing along the bed in a moving stream of chilled air; this is a fluidized-bed freezer and one of the most efficient ways of rapidly freezing particulate foods (and again a perfect analogy for a similar piece of equipment used for drying food, just with much hotter air!).[6]

So, water is a key constituent of almost all foods, and perhaps plays the least showy but most significant role of all. Proteins and their structures might be impressive with tricks like making cheese from milk and turning a raw piece of meat into a succulent steak on cooking, while fats control texture and cause us calorific nervousness and sugars do everything from providing pleasing sweetness to controlling textures of foods from ice cream to fruit, but yet without water almost none of the properties of food we know and expect could exist.

Water provides us with one of the key necessities for life to continue, which is good when this life is ours, but less so when it is that of a pesky bacterium whose presence can cause spoilage, illness, or worse, and so we need to control these factors through processing, where once again, whether cooking, drying, or freezing, we depend on and exploit the properties of water.

---

[6] I always pictured a human version of the fluidized bed, where people, perhaps in giant plastic bubbles, are carried along a track tumbling and bumping in a flow of air, like floating sumo wrestlers. I would very much like to try this, should someone wish to invent it.

One of the simplest molecules in food, almost comically uncomplicated in its single oxygen atom and two bound hydrogen atoms, and yet so much power. Water, we salute you and your quiet power, and your control of the apparently much more complicated molecules with which you share food, deploying impressive and complex structures, but still in many cases dependent on the munificence of your actions for their properties.

{ 10 }

From Napoleon to NASA

THE DEVELOPMENT OF FOOD PROCESSING

Some thoughts on processed foods

One term that has acquired a particular air of consumer suspicion in recent years is "processed food." Processing is seen as being something that is used to make food less fresh, less natural, and so more suspicious.

However, even though we say we don't want processed food, every food product, before it gets to your mouth, has been subjected to some form of processing and treatment that has a scientific basis. Even washing an apple, chilling sushi, or peeling a banana are forms of food processing, while the bag of salad we buy in a shop or market isn't full of air as we might expect. All of these phenomena I will discuss in coming chapters.

Before dealing with the science of food processing, it is worth discussing what exactly that highly loaded term means. To a food scientist (well, me anyway), food processing means subjecting foods or raw materials to external forces designed to cause a desirable change in the food, typically in terms of its safety and stability, and also in many cases its flavor, texture, and color.

In many cases, the primary target of food processing is the resident population of contaminating microorganisms that, if not dealt with, might otherwise cause the food to spoil, or else spoil the day of consumers who finds themselves with a range of symptoms of food poisoning, up to the most lethal. The force most commonly applied in processing is temperature, whether low (refrigeration), very low (freezing), high (for example, pasteurization or cooking), or very high (canning or sterilization).

Temperature is indeed probably the key physical variable of significance to food, as almost everything that happens in and to food is influenced by temperature, and most changes take place optimally in a relatively narrow band around body temperature (37 °C). If temperature is pictured as a line scale, the zone of greatest danger and likelihood is centered around that point, but food processors look far below and above that zone and have come to understand how we can work around the optimum temperatures for various reactions and biological changes in order to make our food safer and more stable.

In fact, this is even pre-science; such knowledge has been innate to mankind for millennia, and we wouldn't even be here today if our ancestors hadn't understood, probably through a grim and corpse-strewn process of trial and error, how to tame nature (surely the first instance of this being done in the history of civilization) in the form of those parts of it that were to form part of our diet.

So, in this book, the term food processing is used to refer to the physical manipulations we apply to food, such as heating, freezing, and drying. Of course, these are not just processes found in large metal vats and other pieces of equipment in food plants, but are also practiced at every scale including the kitchen, and have been so for a very long time.

A French philosopher called Claude Lévi-Strauss described a principle called the Culinary Triangle, in which all food is categorized as Raw, Cooked (processed), or Rotten (spoiled), with Raw often being presented as being at the top of the triangle. Different sides of this triangle then represent different transformations, with the move from Raw to Rotten being due to nature (although this may be favorable, in the case of fermentation, where we have grown accustomed to the "Rotten" version) and the move from Raw to Cooked being due to deliberate transformation, whether cooking or processing.

In recent years, the term "ultra-processed foods" has also appeared. This term is often today applied to food products that contain potentially unhealthy (if they form too high a proportion of our diet) ingredients such as sugar, salt, and fat. The term is also sometimes used in food products that include ingredients not typically used in conventional cooking, like emulsifiers, modified proteins or starches, sweeteners, hydrogenated fats, and colorings (sometimes referred to as ingredients your grandmother would not recognize).

Interestingly, a quick search of scientific papers using the term "ultra-processed foods" finds the description being applied almost uniformly in a nutritional or consumption context, rather than in a food science context. In the context of the health of consumers, these terms are used to classify foods into categories the significance of which for conditions like obesity can then be evaluated and proper nutritional guidelines developed.

As an example of how food products can be classified based on processing and ingredients, the NOVA categorization system developed in Brazil defines food products as the following:

1. Unprocessed or minimally processed, including plants, fruit, meat, water, or products treated physically so as to extend their shelf life, such as by chilling, pasteurizing, drying, or fractionation;
2. Processed culinary ingredients (such as salt, honey, unmodified starches, and edible oils) that have been recovered from natural food products and are used in kitchens as parts of dishes;

FIGURE 2.1 *Nature's perfect packaging solution: the banana (image courtesy of Brian French)*

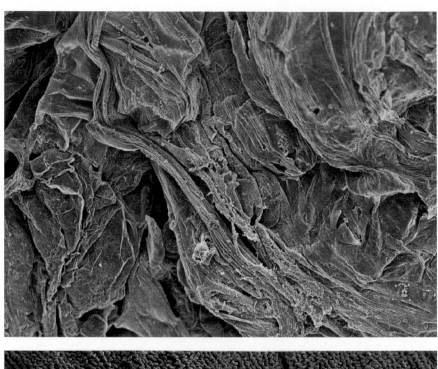

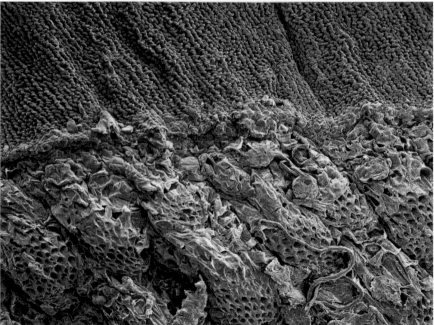

FIGURE 2.2. *A study of bananatomy, showing (from the top) flesh, inner skin (pictures courtesy of Suzanne Crotty, Biosciences Imaging Centre, UCC).*

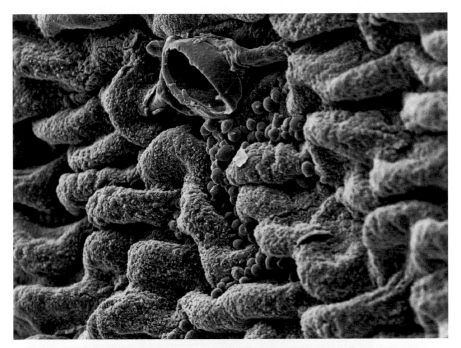

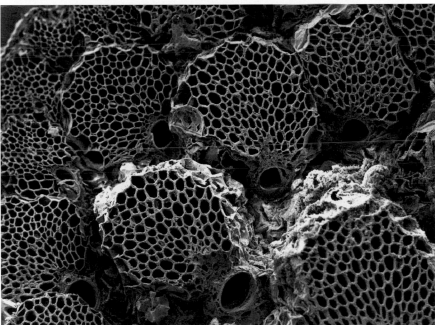

FIGURE 2.2. *Continued outer surface, and stem of a banana*

FIGURE 2.3 *As if designed by an artist: (top) a cut pepper (image courtesy of Brian French), and (bottom) its ghostly inner geometry (picture courtesy of Suzanne Crotty, BioSciences Imaging Centre, UCC).*

FIGURE 2.4 *(top) A tomato (image courtesy of Brian French) and (bottom) its inner microstructure (picture courtesy of Suzanne Crotty, BioSciences Imaging Centre, UCC).*

FIGURE 3.1. *Aged parmesan cheese (image courtesy of Brian French).*

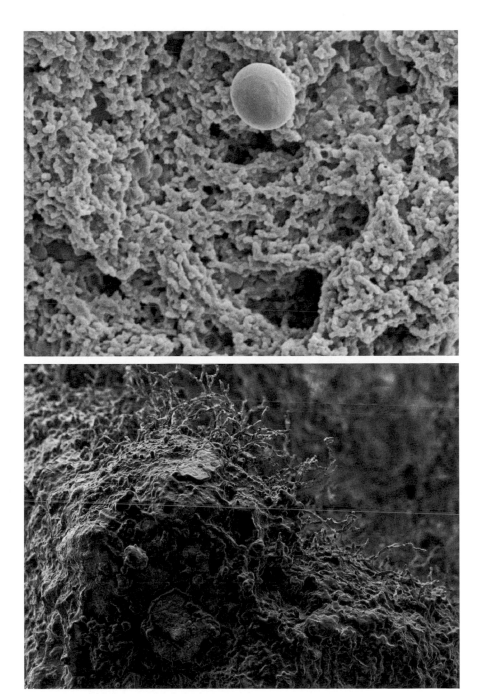

FIGURE 3.2. *The tangled web of the protein–microbe complex in (top) yogurt and (bottom) Blue cheese, with entrapped cells of lactic acid bacteria in the former and tangled fungal hyphae in the cheese (pictures courtesy of Suzanne Crotty, BioSciences Imaging Centre, UCC).*

FIGURE 4.1. *Chicken, in its raw (left) and cooked (right) forms (images courtesy of Brian French).*

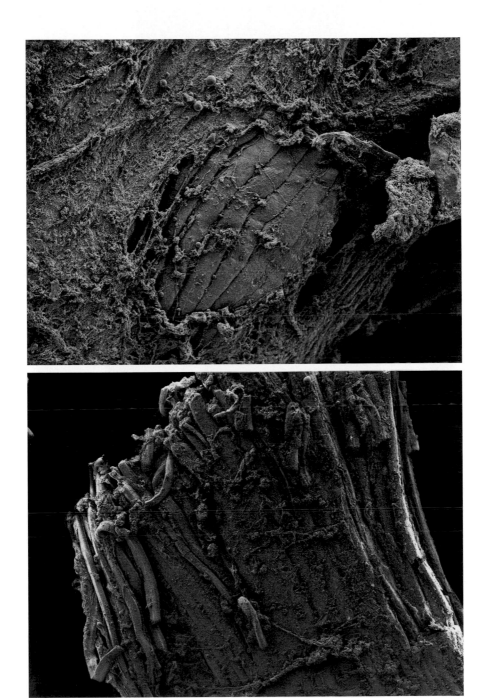

FIGURE 4.2. *The protein-driven transformation of chicken breast from raw (top) to cooked (bottom) (pictures courtesy of Suzanne Crotty, BioSciences Imaging Centre, UCC).*

FIGURE 4.3. *The manifold complexity of an egg (image courtesy of Brian French).*

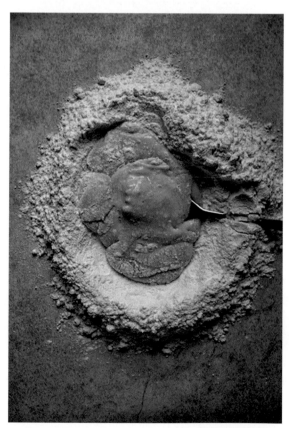

FIGURE 4.4. *Egg yolks being beaten and mixed with tipo 00 flour to create pasta dough (image courtesy of Brian French).*

FIGURE 4.5. *Three views of an egg, from the inside of the shell (top) to the cooked white (middle) and yolk (bottom) parts (picture courtesy of Suzanne Crotty, BioSciences Imaging Centre, UCC).*

(A)

(B)

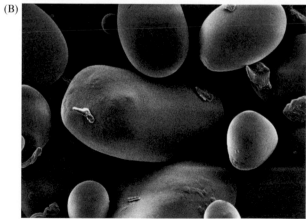

FIGURE 5.1 *(top) The humble potato (image courtesy of Brian French). (bottom) Starch granules in an uncooked potato, showing a remarkably similar appearance to their parent food (picture courtesy of Stefan Horstmann and Suzanne Crotty, Biosciences Imaging Centre, UCC).*

FIGURE 5.2. *A cocktail of hazelnut "milk" and apple brandy that has been spherified with calcium chloride and sodium alginate (image courtesy of Brian French).*

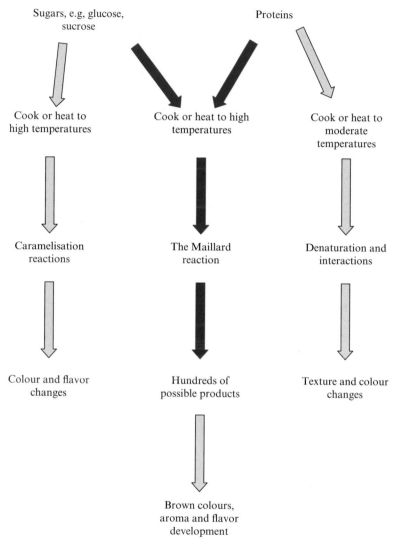

FIGURE 5.3. *A schematic diagram of the complex reactions in which sugars can take part, with or without proteins, when food is heated or cooked.*

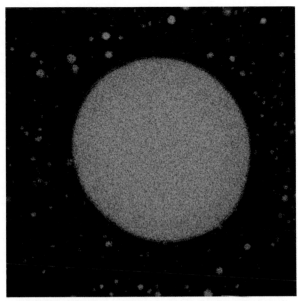

FIGURE 6.1. *A microscopic image of a single droplet in a food emulsion, showing fat (green) surrounded by a thin protein coating (red) (picture courtesy of Dr Kamil Drapala, UCC).*

FIGURE 6.2. *When emulsions break down: a whole (left) and a split (right) mayonnaise (images courtesy of Brian French)*

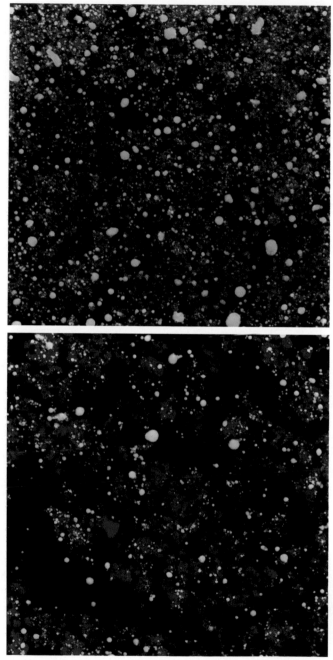

FIGURE 6.3. *Close-up with an emulsion: full-fat (top) and low-fat (bottom) cream cheese samples, showing fat globules (green) in a protein network (red), with water cavities (dark regions) (pictures courtesy of Drs. Juliana De Costa Silva and Seamus O'Mahony, UCC).*

FIGURE 6.4. *A caramel mousse: Maillard reactions meet the foaming properties of proteins (image courtesy of Brian French).*

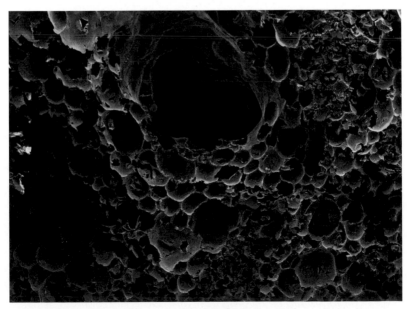

FIGURE 6.5. *Caves within caves: air entrapped in the structure of a meringue (image courtesy of Suzanne Crotty, BioSciences Imaging Centre, UCC).*

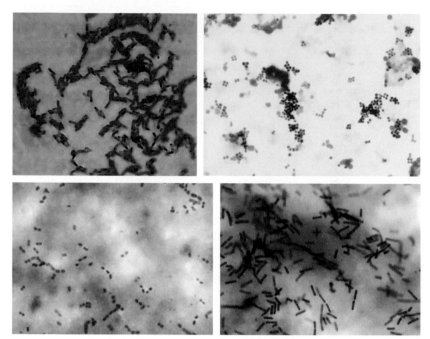

FIGURE 7.1. *Bacteria that may be found in dairy products, showing (top left) Bacillus cereus, which can spoil pasteurized milk (top right), Staphylococcus aureus, a common food-poisoning bacterium, and (bottom left) Lactococcus lactis and (bottom right) Lactobacillus helveticus, both of which are welcome components of cultures used to make cheese (pictures courtesy of Paddy O'Reilly, UCC).*

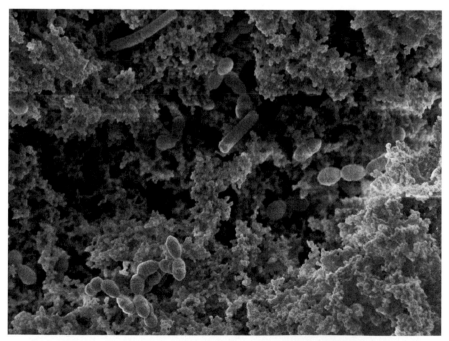

FIGURE 7.2. *Yogurt, showing the rod-shaped cells of Lactobacillus bulgaricus, subspecies delbruckii, and the smaller oval cells of Streptococcus thermophilus in a coagulated casein network (picture courtesy of Suzanne Crotty, BioSciences Imaging Centre, UCC).*

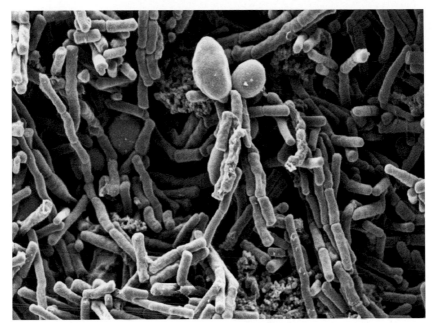

FIGURE 7.3. *The complex microbial ecosystem inside a kefir grain, with yeasts (oval cells), bacteria (long rod-shaped cells), and sections of a protein network (picture courtesy of Loughlin Gethins, Suzanne Crotty and Dr John Morrissey, UCC).*

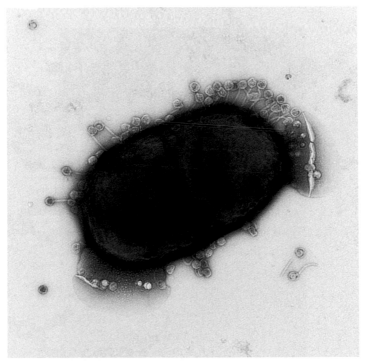

FIGURE 7.4. *Bacteriophage viruses ganging up on a cell of a lactic acid bacterium typical of a cheese or yogurt culture. The tails of the phage have attached to the cell's surface to prepare for injection of their DNA, while the larger, regularly shaped heads point away (picture courtesy of Dr. Jennifer O'Mahony, UCC).*

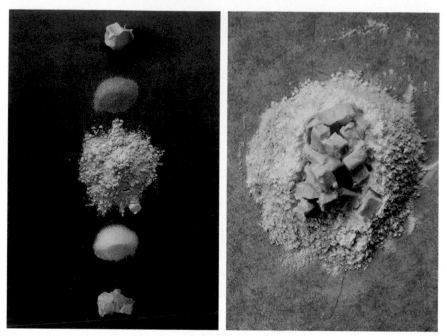

FIGURE 8.1. *The stages of transformation of bread. Part 1. Raw materials of (top to bottom) butter, dried yeast, flour, sugar, and water, separately on the left and mixed on the right (pictures courtesy of Brian French).*

FIGURE 8.2. *The stages of transformation of bread. Part 2. The ingredients mixed into dough (picture courtesy of Brian French).*

FIGURE 8.3. *The stages of transformation of bread. Part 3. The dough rises (picture courtesy of Brian French).*

FIGURE 8.4. *The stages of transformation of bread. Part 4. The final baked bread (picture courtesy of Brian French).*

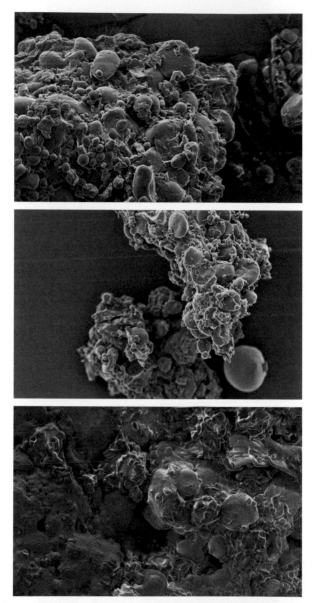

FIGURE 8.5. *The stages of transformation of bread. Part 5. From the top, wheat flour, wheat dough, and baked bread showing the progressive development of structure involving starch granules and adhering proteins, especially gluten (pictures courtesy of Stefan Horstmann and Suzanne Crotty, BioSciences Imaging Centre, UCC).*

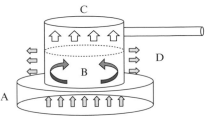

FIGURE 9.1. *The principles of heat exchange that act when we heat a liquid like milk in a saucepan. The hot plate conducts heat across the base of the saucepan (A), which then travels through the milk in the pan by convection (B), while some heat is lost as radiation (D) and water vapor escapes from the surface (C), cooling the liquid as it goes.*

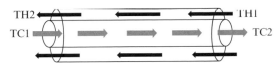

FIGURE 11.1. *The principles of heat exchange that act when we heat a liquid like milk in a pipe in a continuous process. In the tubes, the cold liquid traveling from left to right in the inner tube is heated (temperature TC2 is higher than the starting temperature, TC1) while the liquid in the outer tube is cooled as it surrenders its heat through the wall separating the two (so that TH2 leaves than TH1).*

FIGURE 12.1. *A painter's palette: a collection of dried spices (image courtesy of Brian French).*

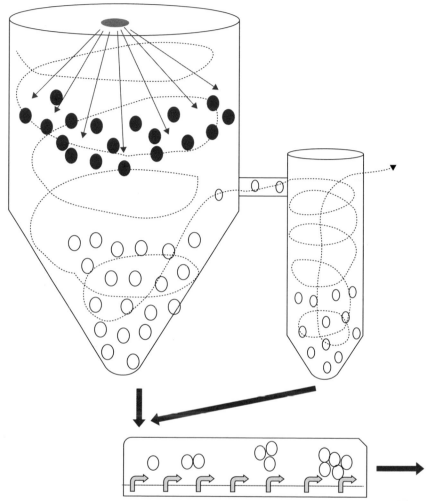

FIGURE 12.2. *A schematic diagram of the operation of a spray-dryer. Droplets are sprayed into the main chamber on the left and meet a swirling cloud of very hot air, leading the droplets to dry into powder particles as they fall through the chamber. When the hot air leaves the chamber, it carries some small particles with it (called fines), which are collected in the cyclone on the right, which acts like the chamber in some modern vacuum cleaners to cause powder to separate from air. The powder from the main chamber and the cyclone then enter the fluidized bed at the bottom, where they are gently dried further, while forming aggregated or agglomerated powder particles, the ability of which to redisperse in water when needed can be carefully controlled.*

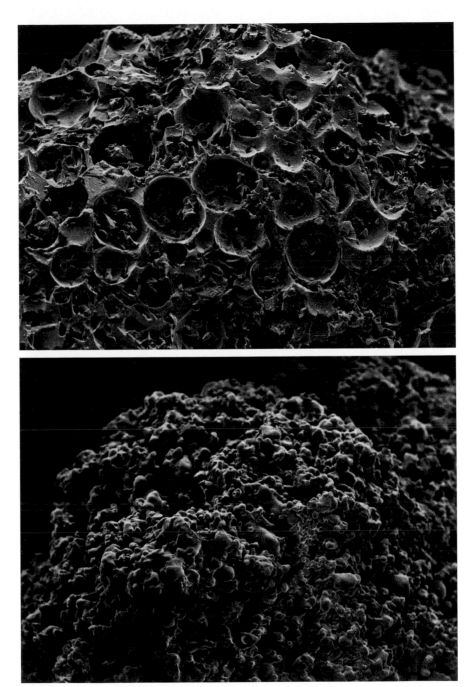

FIGURE 12.3. *The difference in appearance of coffee granules made by freeze- drying (top) or spray- drying (bottom) (pictures courtesy of Suzanne Crotty, UCC).*

FIGURE 12.4. *The geometrically aggressive shapes of granulated sugar crystals (top), and the broken-down structure of icing sugar (bottom), seen at the same magnification (pictures courtesy of Stefan Horstmann and Suzanne Crotty, UCC).*

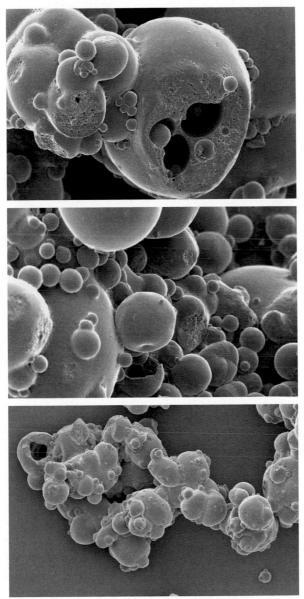

FIGURE 12.5. *Some examples of particle shapes that can be found in dried milk powder (pictures courtesy of Dr. Kamil Drapala and Suzanne Crotty, UCC). The bottom image is taken at a lower degree of magnification than the upper two.*

FIGURE 17.1. *A raviolo (large, single ravioli) that is filled with crabmeat and accompanied with tomato soup that has been ice-filtered to leave a clear liquid (image courtesy of Brian French).*

FIGURE 17.2. *The Maillard reaction in action. A whole celeriac that has been roasted in an oven and basted with honey and thyme (image courtesy of Brian French).*

FIGURE 17.3. *Pea soup with egg and jerked pork (image courtesy of Brian French).*

FIGURE 17.4. *Egg, avocado, and smoked salmon (image courtesy of Brian French).*

FIGURE 17.5. *Mackerel grilled to make the skin crispy, with a burnt apple purée and an apple dashi, a traditional Japanese broth that is made with a selection of dried seaweeds and pickled fennel (image courtesy of Brian French).*

FIGURE 17.6. *Tempered white chocolate and dark chocolate eggs with a coconut panna cotta and a mango and passion fruit fluid gel (image courtesy of Brian French).*

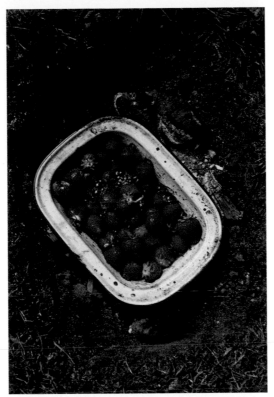

FIGURE 17.7. *Smoked strawberries (image courtesy of Brian French).*

FIGURE 17.8. *A ganache (image courtesy of Brian French).*

FIGURE 17.9. *Salted "bodist" pork ribs that have been marinated in a "jerk" seasoning before being slowly cooked to break down the collagen and make them tender (image courtesy of Brian French).*

Young cheese   Mature cheese

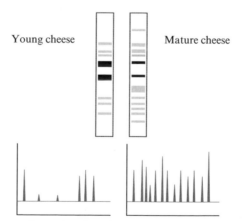

FIGURE A.1. *How we visualize and analyze changes in proteins in a complex system like cheese. In the top schematic images are shown the bar-code-like patterns that are produced when a mixture of proteins is separated by electrophoresis, with each band representing one protein in the mixture. As the cheese in these examples ages from fresh curd to mature cheese, some major (darker) bands disappear as enzymes digest the protein, while other paler bands change, grow, or disappear, reflecting the dynamic environment in the cheese. In the lower figures, corresponding chromatography profiles of extracts from the cheese are shown, and it can be seen that the simple pattern with few peptides (peaks) in the left-hand panel becomes much more complex in the mature cheese profile shown on the right, as the products of the busy enzymes become visible.*

3. Processed foods made by adding compounds in group 2 to products in group 1, such as is the case in salted meat or sweetened fruits, with the goal of increasing shelf life.

4. Ultra-processed foods and drinks (UPFDs), which cover industrially produced products containing five or more ingredients, particularly chemically produced or modified ingredients, or using technologies (for example, extrusion) that have no analogy in a kitchen context. Many convenience foods, carbonated drinks, breakfast cereals, energy bars, and processed meat products fall into this category.

In this approach, processing is defined more by the additives (type or number) that are added to food and their origin than the technologies we apply, with some exceptions, and a red line is seemingly drawn between products that might be produced in a kitchen using familiar materials and methods and those that are essentially made only in an industrial-factory-based context.

The final NOVA category of foods implicitly includes cheap food produced by globalized corporations flooding low- to middle-income consumer diets. UPFDs are also proposed to be hyper-palatable (because of their inducing a sort of sensory-pleasure overdose), making them effectively addictive while providing inadequate intakes of critical micronutrients. The NOVA system is, however, somewhat controversial, even among nutritionists, compared with epidemiological approaches that consider nutrient intake across all food intakes.[1]

The E-number system, which is used to categorize food additives systematically in Europe (hence the E), was developed by the Codex Alimentarius Commission, established again by the United Nations (the Food and Agriculture Organization [FAO] and the World Health Organization [WHO]) in the 1960s to oversee consumer health and food safety. In this system, numbers from E100 to E199 refer to colors; E200 to E299 to preservatives; E300 to E399 to antioxidants and acidity regulators; E400 to E499 to thickeners, emulsifiers, and stabilizers; E500 to E599 to acidity regulators and anticaking agents; E600 to 699 to flavor enhancers (like monosodium glutamate [MSG]), E700 to E799 to antibiotics; and higher numbers to miscellaneous other additives and chemicals added to food. Traditional food preservatives, such as salt, do not have E-numbers, but interestingly carbon dioxide does (E290), as do colors used for centuries, such as saffron (E164) and paprika (E164d).

E-numbers have developed a somewhat negative connotation for many consumers and are sometimes regarded with suspicion, if not actually forming a basis for rejection of a food product. However, they are simply a classification system, in which many components found in unprocessed foods, such as vitamins B1 and C (E101 and E300, respectively), lecithin (E322), and the tomato nutrient lycopene (E160d)

---

[1] As well explored in an article by Gibney et al. in the *American Journal of Clinical Nutrition* in 2017 entitled "Ultra-processed foods in human health: A critical appraisal."

have their own numbers. Notably, E948 is oxygen, which is rather hard to regard as a harmful additive! In addition, the allocation of an E-number indicates that a particular substance has been tested and determined to be safe for use in food, usually within specified limits and terms of use.

While I fully agree with the importance of a healthy and balanced diet (even as I admittedly fail to practice this in my own life!), and similarly cannot argue against the importance of having systems that help to define good practice and on which sensible dietary recommendations are based, I must admit to remaining perturbed by the "kitchen = good, factory = bad" duality that seems to be implicit or explicit in some discussions on this topic.

As we have seen, processing is largely (but not solely) concerned with preservation, and salt and sugar represent at their core chemically based methods of food preservation. Also, many products that fall into the "ultra-processed" group have been produced for decades and offer many advantages to consumers, such as reliability, safety, convenience for a time-pressed lifestyle, and affordability. The key point is perhaps that growing taste preferences may demand their use at levels or in scenarios where their original beneficial function of preservation is outweighed by their unhealthy implications, emphasizing that balance and avoidance of excessive consumption are key principles of a healthy life.

Humans have consumed fat, salt, and sugar in their food for thousands of years and have unquestionably become programmed to accept them and even expect them in their food. Our mouths are full of highly specific receptors tuned just to their flavors. It has been suggested that their taste became associated with safety, in that we instinctively welcomed them as an indication that their presence reassured us that the food we were consuming was unlikely to make us sick.

In the days when preservation of food depended on high levels of salt, sugar, alcohol, or other such substances, they were thus relied on to give safety and stability in the absence of a broader understanding of how to manipulate the chemistry and microbiology of food.[2] Advances in hygienic preparation, storage, and usage of food have rendered most foods less dependent on the role of chemical preservatives. For example, butter has been salted for centuries to preserve it, but butter prepared in a modern facility from pasteurized milk with good quality control and cleaning probably doesn't need the assistance of salt to keep it free from microorganisms,[3] and we continue to have salt in most commercial butter simply because that is what most consumers expect and demand. So, many food additives have a role and have

---

[2] Apparently, in past centuries consumption of alcoholic beverages on a daily basis for most people was vastly higher than it is today simply because that was less dangerous than drinking water because of the presence of antimicrobial alcohol.

[3] Butter is a very safe and stable product because it contains so little of the key ingredient that sustains problematic biology, i.e., water. Butter contains only around 15% water, and that is distributed in tiny droplets, which are an isolated and lonely place for troublesome life to become established, so filling these pools with harsh salt is perhaps only an unnecessary belt-and-braces approach to ensuring biological aridity.

had that role for thousands of years, but today it is clear there can certainly be too much of a good thing.

There is no doubt that a whole book could be devoted to arguments on this topic and that this area is one aspect of food today that arouses very significant passions. The area of processing is the intersection of issues to do with the industrial production of food (and implications for cost, scale, and the environmental consequences of shipping ever-more-stable food around the planet to give distant consumers greater variety in their diet) and the dietary consequences of formulations of low-cost food (such as globally increased obesity related to increased calorific intake). Many books and articles have been written articulating the arguments far more eloquently than it is my ability or intention to do here, given the objectives of this book.[4]

For the rest of this chapter, I am going to leave aside further discussion of nutrition (where I feel myself to be standing on somewhat controversial ground, a little far from the comfort zone of my expertise to espouse wisdom) and focus mainly on what is surely the primary responsibility of anyone who produces food for us to consume, whether in a kitchen or a plant, which is to ensure its safety.

The first topic we consider here is perhaps the oldest known way of reaching this goal, which is heating, and I would also like to consider some key historical steps that have brought the science of food processing to where it is today.

Before that, though, it might be worth pausing to consider what is exactly meant by "shelf life," as already mentioned several times, and in the chapters to come.

What do dates on food packages mean?

One area of information provision to consumers that can be as confusing as it is critically important is the part of the label on the package which gives a date beyond which something will apparently happen to the food product therein that is not desirable.

In the European Union (EU), food packages can present either a "best-before" or "use-by" date. In the United States, the term "expiration date" might be found more commonly than the latter, but it is not legally required for such information to be provided on most food packages (the exception being infant formula), and different states have different rules for this aspect of food labeling. These dates might be specified to a day and month for shorter-life products (shelf life of days or weeks), or simply month and year for longer-shelf-life products, and in all cases refer to products being stored under conditions specified by the manufacturer (if someone keeps a chilled product at room temperature, for example, all bets are off). The difference between best-before and use-by dates is both huge and frequently misunderstood, and perhaps leads to huge amounts of wasted and unnecessarily discarded food.

---

[4] I would refer to one very interesting article here, the title of which gives a clear indication of its position, and mine ("In defence of food science," by McClements, Vega, McBride, and Decker, published in a journal called *Gastronomica* in May 2011).

As discussed throughout this book, while all food products can be susceptible to change after manufacture because of microbial, chemical, physical, or enzymatic processes, some are inherently less likely to become unsafe for consumption as a result than others.

Foods that have conditions that are favorable to the growth of pathogenic microorganisms even after processing, for example, must display in the EU a "use-by" date, to indicate the date after which the manufacturers have determined that there is a risk of harm occurring because of such growth. The product should not be eaten, cooked, or frozen after this date, even if it looks (and smells) fine. Such dates will have been determined by a number of methods, from "spiking" trials including highly controlled studies of survival and growth of the microorganism of concern to mathematical models based on previous studies of similar products. Examples of products that require a use-by date include pasteurized milk, meat, fish, and ready meals.

For some products, there is no expectation of changes during storage that would give rise to a hazardous outcome, and so the main concern might be that changes during storage would result in a less-acceptable product, for example in terms of flavor or appearance. In such cases, a "best-before" date is applied to indicate to consumers that, after that date, the manufacturer will not claim that their ideal eating experience will be delivered. A best-before date will be found on products such as canned food (where the heat treatment has eliminated all microorganisms, pathogenic or not), heat-treated fruit juice (where the highly acidic conditions will make growth of any pathogens that might have survived the heat treatment impossible), or frozen or dried food products (as these processes will essentially have prevented the growth of pathogens).

A best-before date does not indicate that the product should be discarded after that date, and it is quite likely that products will be fit (and safe) for consumption for quite some time after the stated date without noticeable consequences. In addition, consumers will often supplement such advice with their own detailed scientific sensory analysis, involving a quick "sniff" to check the odor of the product and make a judgment call on that basis. Nonetheless, it is unarguable and has been demonstrated in many studies that much food is discarded after the best-before date on the impression that this reflects safety rather than quality.

So, shelf life may be determined by safety (reflected in a use-by date) or quality (reflected in a best-before date). Sometimes, just to help confuse things further, "display-by" or "sell-by" dates might appear on packages, but these are communications from manufacturer to vendor and should not influence the behavior of consumers. In some cases of highly stable products, such as high-alcohol products or highly acidic ones such as vinegar, dates may not actually be required.

Surely a huge amount of food wastage would be avoided if food products that have a best-before date have a second date that would be the recommended last date for consumption and beyond which discarding is recommended.[5] Today, reducing

---

[5] A number of studies in different countries have indicated that at least a third of consumers will not eat food after its best-before date, and many websites now give advice on how long certain food products should be okay to consume after this date.

waste so significantly seems to be a highly desirable goal for many reasons, and some retailers do seem to be taking steps in this regard, with increasing numbers of examples of food beyond its best-before date being sold at discounted prices and regulatory authorities pressing manufacturers to adopt clearer labels and information provision.

A key historical milestone

As mentioned earlier in this book, the contribution of a small number of 19th-century Frenchmen to the development of systematic and controlled heating for the purpose of making food safe cannot be overemphasized. The first, least scientific but most pragmatic, was Napoleon, through his nurturing and celebrating the work of a young chef called Nicholas Appert, who, according to legend, stepped forward to claim a 12,000-franc prize, offered some years earlier, for the development of breakthrough technologies for the preservation of food.

Appert proposed that food could be made safe by placing it in glass containers (originally champagne bottles, pretty far from the cans that are the most common modern method of packaging to follow the same principle), removing as much air as possible and keeping it out with a suitable seal, and then heating for long times (up to 12 hours) in boiling water. His initial preserved products were tested by sending them on French Navy ships on long voyages and, when reports of happy and unpoisoned sailors returned to port (even if, presumably, some of the ships didn't, because of the wars being fought at the time), the method was accepted as a(nother) French revolution.[6] The package used quickly changed to more jar-like containers, and eventually metal (steel or iron) "cans," which, in the 30 years between their invention and that of the can opener, required some serious attacking to open. Perhaps the pinnacle of Appert's achievement came about when he exhibited an entire sheep preserved in a (presumably rather large) can. In recognition of his contribution to the field, the sterilization of canned food was for many years, and even occasionally still is, called "appertisation."[7]

Some decades after Appert, a more recognizable name, Louis Pasteur, began to explore the science behind preservation of food by heat, through understanding what was causing the problems that heat could cure. Pasteur used microscopes to identify and observe bacteria in milk, wine, and beer, and showed that their proliferation accompanied visible spoilage.[8] Pasteur also demonstrated beyond argument that, when microbes appeared in a material, they had to have arrived from

---

[6] Presumably giving Napoleon a new hope that his Empire could strike back against the phantom menace of food poisoning and making Appert a star of the Napoleonic wars.

[7] Not to be confused with appertisers.

[8] The use of microscopy was a key step in understanding how food spoilage was caused, and was a key step because, as we have seen, in terms of resolution, the size of bacteria is far past your eyes to resolve.

somewhere external to that material (as a contaminant, in other words) rather than magically appearing by what was called spontaneous generation, and he contributed hugely to the understanding of microbial fermentation.

Interestingly, his main work on how food could be preserved by milder treatments than those applied by Appert related to the preservation of wine and beer, rather than the milk with which his name has been immortalized in terms of the name of the thermal process applied.[9] Pasteur also made huge advances in the understanding and treatment of diseases such as rabies, and the development of vaccination, but his key contribution to food preservation focused on opening a second front of the thermal war on microorganisms in food, with less-severe consequences for food quality than canning.

We could spend a whole book discussing the history of food processing, the way in which practical experience long preceded scientific understanding, and how in most cases the processes we use today have very long histories. There are also great stories to be found such as the fact that an actual person called Clarence Birdseye (1886–1956; not apparently a captain, except perhaps a captain of industry) was instrumental in the development of processes for freezing of food, based on his observations of how Inuit Indians in Newfoundland achieved excellent quality of thawed fish by flash-freezing (extremely quick freezing giving tiny ice crystals which did little damage to the delicate structure of the fish), which was the consequence of exceptionally cold local temperatures.

However, that is not the purpose of this book, and so, following this brief (but important) historical detour, we now turn to understanding exactly how bacterial numbers are reduced by heating and how being able to predict this allows the modern food processor (and chef) exquisite control over the safety and quality of their products.

The lottery of bacterial death

To understand how heat kills bacteria requires understanding the strange lottery of microbial death behavior. In this regard, it is perhaps lucky (notably not for the bacteria though) that, when heated, they die in mathematically predictable patterns. We also need to appreciate that, when we talk about bacteria growing and dying in food, we need to think about really really big numbers.

In a gram of food, a million bacterial cells would not be a surprising level to find, and a billion (a thousand million or 10^9) wouldn't be unusual (these are really big numbers, if you start thinking of these as populations of countries, or maybe bank balances). If we heat that food for a particular time, the levels of surviving bacteria can be predicted quite easily, as they die in a particular pattern called an exponential rate.

---

[9] The term pasteurization is used today to refer to a treatment sufficient to kill vegetative (non-spore-forming) bacteria in a whole range of foods, from beer to eggs.

Let's start with a hypothetical food system with 10 million bacteria present in each gram of product. If, in 1 minute at a particular temperature, 90% of that starting population dies, this leaves one million cells. We call such a reduction a one-log-cycle reduction (a 1-in-10 reduction). Then, in the next minute at that temperature, another 90% will die, leaving a hundred thousand cells, and in the next minute we will reach ten thousand. On this basis, we know that if we heat for more than seven minutes we should be left with fewer than one cell in that gram of food.

The time it takes to reduce the number of bacteria by 90% (or a factor of 10) on heating at a particular temperature is called the D-value (short for the decimal reduction time). The higher the temperature, the less time to achieve a certain level of killing, as you might predict, and so the lower the D-value.

To look at it in a more familiar way, for the bacteria being heated, the D-value means the odds of their survival. Heat a sample of bacteria for one D-value's worth of time, and the chances of survival of any of them are 1 in 10. For every 100 cells, 10 will survive. If we heat for four times the D-value time, then the chances are 1 in 10,000 (as 1 D is 1 in 10, 2D is 1 in 100, 3D is 1 in 1000), and the heating will kill 99.99% of bacteria. If we started with 1000 bacteria, it is unlikely we have any survivors after four D-values' worth of heating, but if we start with a million we might have a hundred left behind.

We frequently encounter advertisements for cleaning products that kill 99% of bacteria. Perhaps it might be imagined that there are 100 types of bacteria, and it kills all but one, but in scientific terms this means that the product can deliver a two-log-cycle reduction, which frankly isn't much to get excited about when we probably start with much larger numbers of bacteria that need to be dealt with.[10]

In terms of the heat-induced killing, a treatment that didn't achieve at least five D-values or log-cycles worth of killing a bacterium we wanted to remove as a threat from our food wouldn't be regarded as very effective, as it wouldn't take care of a typical level of bacteria we might expect to be there to start with (as having 10,000 cells in our food wouldn't be at all unusual).

If we want to make food sterile and ensure that there is no risk of growth that could possibly lead to health or spoilage problems, we need to consider first the toughest bacterium we might encounter, and so that we might need to be sure we can kill. For many years, this was regarded as a bacterium called *Clostridium botulinum*, which is a tough spore-former found in soil or dirt, and that can produce a highly lethal toxin that leads on consumption to paralysis and death, in the nightmarish syndrome called botulism.[11] *Clostridium botulinum* is often used as the benchmark target for inactivation of bacteria in food as, if this is the toughest

[10] I recently saw a sign at an outdoor music festival for hand lotion that was an "anti-bactericidal gel." This is the equivalent of anti-homicidal and is surely not what was meant.

[11] Food scientists work diligently to protect consumers from this terrible risk. Despite this, some consumers pay large sums of money to have that bacterium's deadly toxin injected into their faces to paralyze their skin and stretch it out, ironing out wrinkles, as this is the origin of Botox.

bacterium present, then heating conditions that were sufficiently harsh to kill it should kill off all other less-tough species.

What would be a level of killing[12] that was sufficient to regard a food product as sterile in terms of *Clostridia*? Let's turn that around and ask what would be acceptable odds of a single cell of *Clostridium botulinum* surviving to produce its awful toxin in a product such as a can of food stored at room temperature for very long periods. In food processing terms, that level of confidence is expressed as 1 in 10 to the power of 12, or 1 in 1,000,000,000,000 (which makes winning the lottery look like a sure thing). This means that, to have a chance of finding one surviving cell, there would need to be at least that ginormous number of bacteria present to begin with, which is, to put it mildly, extremely, even astronomically, unlikely[13].

So, heat a food product for 12 D-values worth of time and the chances of a spore surviving are so remote that we can assume it is sterile. As it happens, careful measurement of one key D-value of *Clostridium botulinum* showed that it takes around 15 seconds to reduce its numbers by 90% at 121.1 °C (a temperature selected a long time ago as a reference temperature when making these kinds of calculations). So, to achieve 12 times this, we need to heat food at that temperature for around three minutes (12 × 15 seconds).

This then becomes a hugely significant benchmark and target in the heating of food with the goal of sterilizing it, and some not-too-complex mathematics[14] can show that the heating conditions applied when we heat food in cans at around 115 °C for 10–20 minutes or UHT-treat milk or other fluids at 138–140 °C for a few seconds[15] are actually exactly equivalent to this benchmark in terms of their ability to kill spore-formers, which explains where these food processing conditions come from. In other words, they come directly from a deep understanding of exactly what are the minimum conditions needed to achieve a particular desirable outcome, and processes are then calibrated to achieve this requirement absolutely, and nothing more severe than that.

Plots of the conditions (combinations of time and temperature) needed to achieve these end results can also be created highly accurately, so that processors can judge exactly what time might be needed at higher or lower temperatures to get the same effect.

---

[12] The term inactivation is frequently used to refer to the killing of bacteria, but I think killing describes more frankly what happens when we apply such processes, as there is no other living species for which we would refer to the transition from a living to a dead state as inactivation.

[13] Ten to the power of 12 is actually not too far off the estimates for the number of stars in our galaxy, the Milky Way, and so the chances of finding other life in that galaxy (to the best of our knowledge) are pretty much this level of unlikeliness!

[14] Which nonetheless I won't get into here, thankfully!

[15] The nature of the equations involved means that the rate of killing of the bacteria increases incredibly fast (technically exponentially) above the selected reference temperature, in this case 121.1 °C, and so an effect that needs minutes below this temperature can require only seconds above it to get the same end result.

This is particularly significant because every second at such high temperatures results in the risk of damage to the nutritional and sensory attributes of food that consumers value, and so overtreatment (or overkill) is neither needed nor warranted. So, selecting conditions for thermal processing often involves a fine balancing act between achieving desirable effects (microbial kill) while minimizing undesirable ones (related to damaging food quality).

Helpfully, the exact same principles previously outlined can also be used to determine the limits of changes in food, like destruction of a particular vitamin or nutrient. For a given product, the conditions (time and temperature) above which more than 10% of vitamin C in a product will be destroyed can be mapped, and conditions to prevent such deterioration selected for use. Thus, careful consideration of these types of data for a product can allow the selection of heating conditions that maximize the desired change while minimizing the undesirable ones.

The kind of calculation previously considered is based around the kind of conditions needed to achieve absolute sterility, like when food is heated in a can. A very slightly modified approach, using different parameters for the bacteria of interest and the reference temperatures included in the equations, allows us to use it to compare the effectiveness of processes at different kinds of temperature, like, say, the effects of heating vegetables in boiling water to 100 °C, or pasteurizing fruit juice at temperatures between 65 and 75 °C (shorter times being needed at higher temperatures).

So, when heat treatment conditions for food products are being designed and recommended for use, these are not plucked out of the air, but carefully selected based on calculations of how to make absolutely sure the product is safe and as stable as we need it to be, while retaining as much of the desirable characteristics as possible.

Now, having focused on how to very scientifically kill bacteria, we should for balance again mention their more positive aspects in the world of food science, for example, in the ancient art of fermentation by which they have, for millennia, been put to productive use making a wide range of desirable food products.

As mentioned several times already, many food products are the result of some kind of spoilage we came to like over the years, and we have learned to exploit some weaknesses of raw materials to give us some of our favorite products. These include lots of examples where we use microbes, which usually we think of as the enemy. However, if it weren't for yeasts, bacteria, and fungi, we wouldn't have cheese, bread, beer, wine, Camembert, and many more products, as discussed in other chapters of this book.

The idea that some phenomena are ideal in one food context but undesirable in others doesn't just relate to microbiological growth. When we cut or damage fruit, we expose their innards to oxygen, which can lead rapidly, if not controlled, to softening and browning thanks to the action of an enzyme called polyphenol oxidase. In the production of tea, however, the same enzyme and family of reactions are critical for the development of desirable colors and flavors, and the action of the enzyme is specifically allowed to proceed during a fermentation step. Likewise, for most food products we want fat, if present, to be neatly trapped within some form of emulsion (as discussed in Chapter 6), as otherwise it would look greasy and shiny, but this is exactly what we want in chocolate, where so-called "free" fat is actually a desirable component.

Hurdles and trampolines: The Olympics of food preservation

Food preservation throughout history has often relied on the "big gesture" model of food processing, where one big change was applied to a food because it was found to make that particular food safe and last longer, and that was done to occasional excess as a result. So, some foods were dried, some were heated, and others were salted, pickled, or fermented. In some cases, combinations of treatments were used, whereby foods were pickled and dried, or salted and dried (like fish or meat), or even fermented and salted (like cheese), and, over millennia and centuries, products we are very familiar with today arose from such intuitive strategies.

In recent years, a critical concept for understanding food safety revolves around understanding how combinations of processes or treatments work together, and recognizing that, if we apply a lot of different processes together or in sequence, we can get synergistic effects that mean that the severity of any individual process or change can be lessened. This has then major implications for being able to change the sensory or nutritional quality of food less through processing, and leave it more like the natural raw materials, in line with the consumer demands and drivers mentioned earlier.

The scientific term for this idea, borrowed perhaps surprisingly from the world of athletics, is hurdles, and this can be best thought of as an organizing concept for how food scientists think about making a food product safe for consumers. This concept was first developed in the 1970s by a German food scientist called Lothar Leistner, and the idea behind hurdles is that every food product possesses a number of factors that in some way oppose the growth of undesirable microorganisms.

These may be characteristics of the product themselves, such as the level of acidity, salt, or alcohol present, the process to which we have subjected the food (such as the temperature to which it has been heated), or the properties of the environment in which the food is stored (such as the temperature, gas composition, or humidity). All of these factors influence the potential growth of microorganisms. At some levels, their effect might be negligible (for example, a tiny level of sugar will have no meaningful preservative effect), while at others their effect might be undesirable in terms of the food (a level of sugar might be so high that the food is unpalatably sickly sweet). In between, then, is a zone where the level suits the food's flavor and at the same time works effectively to inhibit undesirable microorganisms. In the case of sugar levels, we could term this the sweet spot.

The term then used for these factors is hurdles, in an analogy with the wooden barriers that can be placed at regular intervals along a running track that athletes must leap to clear before progressing. Picture a really long track, with lots and lots of hurdles, perhaps increasing in height as the track proceeds. Perversely, the objective of this race is not to see who can get through the track at all, but rather to design a track that no athlete can complete because of its toughness. A rather weak and unfit athlete might clear the first few hurdles okay, but with increasing difficulty, and, on being confronted with the next, simply collapse, unable to go on. The most fit might get much farther, but eventually arrive at a devilish hurdle that defeats their flagging resources of energy and will to fight on, and there they fail.[16]

What is unfair torture for athletes, however, is the basis of how to design a food product and process to prevent growth of undesirable microorganisms. The idea is that each of the factors mentioned earlier represents a hurdle, and the combination of hurdles should collectively represent a challenge too great for survival or growth to dangerous effect of the microorganisms of concern. Modern food product or process design, perhaps always implicitly but now more explicitly, uses the principle of the hurdles to guide the design of process and formulation, the objective being to balance all the potential preservative factors (like acidity, heat, cold storage, and the presence of antimicrobial substances) so that the net effect is decidedly hostile to unwanted growth and change.

The principle underpinning this concept is called homeostasis, which refers to the ability of organisms (such as bacteria) to adapt to stresses to maintain their survival. If we hit bacteria with a single stress (heat, acid, etc.), there is a chance some may adapt to survive the challenge, but, if we hit them from multiple angles at once, they have far less ability to cope. Many military leaders have commented through history of the ill-advisedness of fighting a war on two fronts at once. Bacteria share the same weakness, and hurdle approaches to preservation exploit this mercilessly.

Designing a preservation strategy for a food product so firstly requires knowledge of the bacteria that might be most likely to be of concern in terms of safety. In the case of eggs, for example, this might be *Salmonella*, while for minced meat burgers it might be *Escherichia coli*. In either case, to make safe a product containing these ingredients requires knowledge of the weaknesses of these key target bacteria, such as a detailed understanding of the temperatures they can withstand or prefer for growth, their resistance to acidity, the maximum levels of salt they can withstand, and other such vulnerabilities. The ideal process for the product in question then takes this insider information and turns it into a set of parameters that are applied in the formulation, processing, or storage of the product.

In the case of a minced meat burger, we also need to think about how the product is produced. In most meat products, the key point of contamination and growth of bacteria is likely to be the surface, which is of course the easiest to decontaminate by cooking, while bacteria are less likely to be found deep within the meat. In making a burger or similar product, however, the original meat has been minced

---

[16] I am not sure whether this would ever be adopted as a sporting event, but I personally think it could make some entertaining television as a new extreme sports challenge, and I am certain some reality TV company is considering it right now.

and reassembled into a patty (making what is called a comminuted meat product) without regard to the starting structure. The surface and associated bacteria may well end up in the hardest-to-heat center, and so cooking right through is even more critical than for other types of meat.

There are also factors associated with any food product that have an effect that is basically opposite to that of hurdles, by actively promoting the growth of unwarranted microorganisms, such as the presence of particular nutrients. These of course have a name complementary to that of hurdles. What would help our poor hypothetical athletes from earlier when confronted with increasingly monstrous obstacles to clear? Because "pogo stick" might have not sounded sufficiently scientific, the only slightly better term "trampolines" is used instead. Hurdles and trampolines; this is how food is preserved today.

To take an example, when we think of how pasteurized milk is preserved, we tend to think of the heat treatment to which it has been subjected, at 72–74 °C for 15–30 seconds. However, there are two additional key hurdles that are critical in ensuring that the product reaches the consumer in a state of optimal quality and safety. The first is maintenance of the cold chain for the product, whereby it is never allowed to warm up after heat treatment to temperatures at which the bacteria (particularly those pesky spore-formers) that survive the heating could start to grow and exert nasty effects, while the second is the protection of the milk in a suitable package that protects it from contamination after the heat treatment.

Pasteurized milk is not sterile, but the heat treatment is sufficient to kill the pathogens (other than spore-forming pathogens that are controlled by refrigeration) and reduce the levels of other bacteria that grow in the cold enough that we get a refrigerated shelf life of maybe two weeks. Such bacteria will eventually start to grow and result in the curdling or spoilage we detect when milk has gone off.[17]

Take pasteurized milk, open the container, and leave it stand on a table for a while, and all bets are off. The good of the heating will rapidly be undone, as the bacteria within or that have got in from the environment begin to bounce happily on the many nutritional trampolines within. So, in this case, three hurdles are critical for making sure consumers aren't made sick by pasteurized milk: heating, cooling and cold storage, and packaging.

In the case of the much more severe UHT treatment, we have increased the height of the heating hurdle so much that bacteria die just at the sight of it, the

---

[17] A common bacterial species responsible for spoilage in many cases is called *Bacillus cereus*, shortened to *B. cereus* and pronounced in a way which could result in the response "I am, this is a major problem for spoilage of pasteurized milk!" Other bacteria that might contaminate pasteurized milk can cause defects like ropiness (rare strains that produce gummy polysaccharides in milk), acidification (if the bacteria present ferment milk), or clearing of the milk from white to translucent (if they produce enzymes that break down the milk proteins). Many of the increases in shelf life and quality of pasteurized milk in recent decades actually arise from improvements in packaging (such as a move from glass bottles to Tetra Pak containers in many countries) that have reduced the risk of what is called post-pasteurization contamination between heating and eating.

upper bar presumably lost in the clouds above, and once we maintain a sealed sterile package we don't need to add the cold storage hurdle.

To add to the complications, hurdles are not static. To extend our extreme sports analogy, imagine if someone came onto the course during the race and adjusted the height of the hurdles or took away a previously perfectly positioned trampoline. This often is what happens in food products as they are stored, reflecting the fact that they are dynamic and changing systems. In the case of a fermented sausage, for example, during storage it dries out (increasing the low-water hurdle), the sugar present is used up by fermentation (reducing the bounciness of the sugar trampoline), and the acidity increases because of said fermentation (increasing that hurdle). So, in this case, the hurdles are different at the end of the product storage than at the start.

NASA and HACCP

It is probably fair to say that being in a space suit far not only from earth but from the spaceship to which you are tenuously tethered is possibly the ultimate example of a time when you don't want food-induced unwellness to strike. A rumbling tummy and some gaseous eruptions from any part of the body would be enough to strike fear into any astronaut and create quite a problem for Houston. There are other particular considerations for such food too, including that it wouldn't crumble too easily into small fragments that would float around in the absence of gravity until, in the zero-gravity equivalent of dropped toast always falling buttered side down, it inevitably lodged in the least suitable piece of electronics or equipment.

For this reason, it is not surprising that NASA, from the earliest days of the space program, was very concerned about food safety. This work commenced in the late 1950s and arose through a collaboration between NASA and the Pillsbury Corporation, who were working on cube-shaped food products intended for space flight and built on a concept borrowed from traditional engineering strengths in NASA, which was defining critical control points for a process.[18] This led to the development of what is called HACCP (Hazard Analysis of Critical Control Points, pronounced Hassop[19]), which today represents the core principle of how food production everywhere from the largest factory to the smallest takeaway (carryout) restaurant is operated (hopefully) in a safe way. The spread of good practice in HACCP from NASA to the broader food industry proceeded through the 1960s and 1970s, spurred by a few high-profile serious cases involving contamination of baby food with glass and botulism arising from underprocessed canned goods.

---

[18] Apparently first introduced for quality control for artillery shell production, so its use in ensuring safety of, among many other food products, shellfish and eggs today is another example of turning swords into ploughshares.

[19] I am positive someone at that company wanted to call it NASAP, but got overruled.

The key to HACCP is depicting a food process in a logical systematic way, perhaps using a flowchart with arrows between steps, and identifying all the steps which are undertaken from the start to finish. This might start with the intake of raw materials and their storage, followed by mixing, heating, cooling, and packaging, and finally dispatch or heating and serving to customers.

Then, for each step in the process, questions are systematically asked about whether that step represents a key opportunity to control food safety if operated correctly. For example, in pasteurizing milk or juice, the key safety-ensuring step is the heating to 72 °C for 15 seconds, or close to this, and this is the bit that absolutely needs to be done right for consumer protection.

This is then labeled a Critical Control Point, as it is a Critical Point at which Control (of Hazards), is exerted, according to the Analysis performed (hence HACCP). Focus then zooms in on this step, as the conditions that are needed for it to have this hugely important effect are defined (exactly what temperature needs to be reached, and for how long, as may for example also be defined for the center of a burger being barbecued on a summer's day). Personnel are then trained on the importance of achieving these conditions, and means for ensuring that they have been reached are defined, so that quality can be assured.

This is the principle of a HACCP plan, and evidence of same can normally be found everywhere food is produced, served, or sold. The HACCP principle has more recently been extended to related offshoots like threat assessment critical control points (TACCP) and vulnerability assessment critical control points (VACCP), which are more focused on the prevention of adulteration of food, undeclared substitution of materials (like horsemeat entering supposedly beef-based products), or food fraud. HACCP was designed to prevent against unacceptable behavior by the unseen microscopic villains that share our world, whereas the two more recent versions are focused on misbehavior by humans.

The ancient art of chemical warfare for food preservation

One term that perhaps more than any other is likely to make consumers anxious is "chemical preservatives" and, as a result, the absence of chemical preservatives can be a major factor influencing decisions to buy or not to buy a product. Manufacturers will even, in recognition of this, draw attention to their products' lack of such evil additives.

However, the addition of chemical preservatives to food is, as I have said, an ancient art, and many common food components we take for granted are there, and have been there for a very long time, specifically because (and there is no way to sugarcoat[20] this) they are very powerful chemical preservatives! In this category,

---

[20] Sugar-coating being, itself, an example of a chemical preservation strategy.

I include such staples as salt, sugar, vinegar, and, above all else alcohol. The traditional method of smoking food is also a carefully designed chemical warfare assault on the microbial population of food, as bacteria are subjected to a stream of substances that to them act like nerve gas or worse.

Consider the following list of molecules: Fluorene, Phenanthrene, Anthracene, Fluoranthene, Pyrene, Benzo(a)anthracene, Cyclopenta pyrene, Chrysene, Benzo(b) fluoranthene, Benzo(a)pyrene, and Benzo(g,h,i)perylene.

Now imagine picking up a product in the supermarket and seeing those on the label. Would that look natural or enticing? Unless you are an organic chemist with an unusual affinity for bringing your work home, I suspect not.

What about this list? Trichloromethane, Methylbenzene (toluene), Alpha- 1,4-Dimethylbenzene (p-xylene), Ethylbenzene, Beta-pinene, Ethyl-cyclopentane, 1,3,5-Trimethylbenzene, 1-Ethyl-2-methyl-benzene, Ethyl-3-methyl-benzene, 1,2-Dimethylbenzene, 2-Methylfuran, Benzofuran, Pyridine.

Sound enticing? Mouth-watering?

The first list I got from a research paper on the substances present in smoked salmon, the latter from a paper on smoked cheese, but in neither case will the product label contain any of these undeniably scary-sounding chemical tongue-twisters,[21] but just the simple, homely, and (again) traditional-sounding one: "smoked."

There should be several conclusions from this consideration:

1. Just because something has been done for centuries does not mean it is not an incredibly complex chemical process;
2. Just because our ancestors didn't understand the fact that they were applying chemical preservation doesn't mean they didn't understand that, if they did this, they would have safe and stable food, which tastes nice, while their graveyards were full of people who didn't do it;
3. Just because a chemical looks unpronounceably complex and scary doesn't mean that, if present at measurable but very low levels, it won't taste nice at Christmas on a thin sliver of fish laid carefully on a piece of buttered brown bread.

Smoke is produced by oxidative combustion (burning) of the wood, which generates a complex mixture of vapors and particulates that get caught in the vapor generated by burning and go along for the ride. Molecules in the wood such as lignins and cellulose are broken down to much simpler ones like acids, carbonyls, and phenols, which are then transported in the gaseous smoke to the surface of the food. Acids give tart flavors, while phenols are fat-soluble and confer flavors to this part of the food, and carbonyls will react with proteins by the Maillard reactions, leading to color development. Different starting proportions of raw materials in different kinds of wood lead to a different buffet of compounds produced, which is

---

[21] Some of which may have astringent tastes, giving a puckering sensation in the mouth, and so being actual tongue-twisters and not just linguistic ones.

why oak-smoked might taste different to beech-smoked and so forth. Hardwoods such as apple, beech, and hickory will give a sweet, clean smoke compared with softer woods, because of differences in particular compounds present.

Smoking impregnates the surface of a food product with low levels of the aforementioned compounds and many more besides, and they collectively and individually exert a wide range of effects, some of them attacking bacteria already present at the surface, or those that land optimistically from the atmosphere onto what they don't realize is actually the chemical equivalent of quicksand. Some of the compounds in the smoke are volatile and escape readily into the atmosphere around the food to create a smoky aroma, while others confer color or flavor. Smoking thus confers character to the food while at the same being a very powerful preservation tool, especially when combined with the loss of moisture that often results from the heat associated with many smoking processes, which reduces further the attractiveness of the environment for microbial growth.

Traditional is not the opposite of chemical! For thousands of years, we have added chemicals to food, like sugar to jam or salt to meat, or else exploited microbial processes that produced them in the food, like fermentation to produce acid or alcohol. In every such case, the reason these have persisted is because of the power of their chemical preservative effects (and, just maybe, in the case of some of them, because of some other popular side effects of that products of fermentation).

We have become socially accustomed to the flavor of many powerful preservative agents, and perhaps their other functions in food systems. For example, as we have encountered several times in the book, salt (sodium chloride) is a highly versatile molecule, to which we have our own dedicated set of taste buds but that has a particularly powerful effect on water in cells in plant- or animal-derived food, or even our own or microbial cells, by essentially being able to direct the flow and behavior of water. If a bacterial cell finds itself surrounded by a high concentration of salt, water will be inexorably drawn out of the cell as nature tries to override what it sees as an unacceptable imbalance of salt levels outside and inside the cell, which it strongly feels must be equalized. The hapless cell then has its life-sustaining water sucked out, and dies as a result, which is the reason for the power of salt as a preservative.

In addition, if the cell in question is in meat, the water-mobilizing power of salt can have major effects on properties like juiciness and tenderness of the eventual meat. When salt meets the outside of meat or fish, nature tries to equalize the difference in salinity on either side of cell walls but, as the cells cannot transport all the salt inward, the only solution is for the water to leave the cell, and, as a result, the cell dehydrates and dies, with obvious consequences either for the meat structure or bacterial survival, depending on which type of cell we are taking about.

In the case of sugar, the preservative effect is likewise related to an ability to deny cells of vital water, dehydrating them into a useless state of desiccation, while giving a pleasing sweet flavor.[22] Acids, like vinegar in pickling or lactic acid in yogurt

---

[22] It is interesting to wonder why so many of our taste buds react specifically to preservative substances, including sugar (sweet), acid (sour), and salt (em, salty)—could this be an evolutionary outcome of our survival as a species thanks to their inclusion in otherwise hostile food products?

TABLE 10.1. A Rogue's Gallery of some food-poisoning bacteria, and how best to kill them

Bacterium	Crime	Weaknesses	Solution
Clostridium botulinum	Paralysis from contaminated canned food	Doesn't like acidic environments or low temperatures	Heat to kill, refrigerate to control, or have a low pH
Listeria monocytogenes	Death from contaminated dairy products, even if refrigerated; pregnant women particularly at risk	Killed by pasteurization of milk	Avoid dairy products made from raw milk if pregnant
Bacillus cereus	Food poisoning, leading to vomiting or diarrhea	Most likely to occur during storage of cooked food, doesn't like acidic or dry conditions	Refrigerate cooked food quickly, or ensure pH or moisture content inhibit its growth
Campylobacter	Severe diarrhea from consuming contaminated poultry	Doesn't like salty conditions, not too heat resistant	Brine meat and make sure turkey cooked above critical killing temperature
Salmonella	Severe illness or death from contaminated food, including eggs	Killed by mild cooking	Avoid undercooked products, especially eggs
Verotoxigenic *Escherichia coli*	Possible death from undercooked meat	Dies at temperatures at which meat changes visibly	Make sure burgers cooked until all pink color disappears
Vibrio	Transmits cholera through contaminated water or oysters	Killed by mild cooking	Avoid raw shellfish

fermentation, reduce the pH of a food product to values where bacteria cannot abide the acidity and lose the will to harm.

Once again, these are all examples of complex chemical phenomena, and just because they have been in use for centuries doesn't mean that they are not chemical preservatives in the clearest definition of the term. Put another way, if salt or alcohol didn't exist already, but were discovered in a lab somewhere and proposed to be widely added to food for their preservative (and many other) effects, would they be allowed today?

Putting it all together: Making food safe through science

To summarize, the main themes of this chapter, the key to making food safe, and also achieving the second goal of processing, which is to make it last longer in an optimal state, are understanding the main microorganisms present that are of concern (usually, but not exclusively, bacteria) and then designing tailored assault strategies to ensure, if not their death, then their rendering utterly harmless.

It is a question of finding their weak spots, and attacking these without mercy, just as any good military strategy will endorse. To show some examples, let's consider what might be the bacterial equivalent of the FBI's Most Wanted[23] list, and what we know about how to kill them (Table 10.1).

The second column of this table is a depressing reminder of how much can commonly be encountered in food that can cause death, misery, or long-term debilitating illness. The third and fourth columns, however, give plenty of cause for hope, showing how the understanding previously mentioned has been turned into good advice for everyone, from home to factory, to ensure food safety. This can be seen to inform advice of all sorts, from how a turkey is cooked to what foods to avoid under certain conditions, as well as good practices like keeping cooked and raw meats separate to avoid cross-contamination.

Yes, it is a scary fact that the world is full of things we cannot see that seem to be designed to cause us horror, particularly when we deliver them directly and highly efficiently to our vulnerable insides by the superefficient method of shoveling them in with our food. On the other hand, thanks to thousands of years of experience and a few smart Frenchmen with prepared minds (and many other food scientists around the world since), we know pretty well how to make sure the nefarious plans of the microbial underworld come to nothing.

In the next few chapters, I will take a few of the principles introduced here, heating, cooling, and drying, for example, and expand on how these are applied in practical terms, in scenarios from the kitchen to the factory.

---

[23] In this case, perhaps Food Bacterial Invaders: Most Unwanted. Now there's another idea for a primetime TV show!

11

Heating and Cooling of Food

SIMPLE, CRUDE, BUT EFFECTIVE

Get it in fast, get it out fast: Optimizing food heating and cooling

As we have seen, heating of food is one of the oldest and most powerful ways of making food safe and stable, whether cooking a burger on a barbecue or pasteurizing juice, but is also a potentially highly damaging thing to do to many food products. So, it makes sense that a key principle of processing food is to understand how to control the flow of heat as precisely as possible.

In Chapter 8, I introduced how we can maximize the efficient transfer of heat into and out of food in a kitchen in simple systems, like pots on stoves. In practice, in large-scale processes, to transfer heat efficiently from hot to cold, and in this way keep the lords of thermodynamics happy while minimizing damage to the food being heated, we need to use clever pieces of equipment, called heat exchangers (reflecting the fact that, just as the cold part of the system gets hotter, so the hot part gets colder in the deal; fair exchange is no robbery).

To visualize a heat exchanger, imagine a simple metal tube, through which a cold liquid is flowing from one end to the other. Now surround that tube with a larger one, through which a hot liquid flows (as shown in Figure 11.1). The wall of the inner tube is exposed to cold on the inside and hot on the outside, and this temperature gradient is the pump that transfers heat across that wall, in nature's obsessive quest for equality in all things temperature-y. So, now we have two tubes laid horizontally in concentric neatness, say with a hot and a cold liquid flowing in from the left-hand side; as they exit at the right-hand side, the outer hot liquid will be colder, and the inner liquid will have gained the lost heat and thus become hotter. If the tubes were sufficiently long, then both would come out at exactly the same temperature.

Now, how do we maximize the heat transfer between these fluids as they pass through their tubes? There are a few basic principles that will help a lot. The first, as mentioned already, is to maximize the difference in temperature between fluids, giving in effect a bigger pump to drive the heat between them.

The second will be to have as unobtrusive a barrier as possible between the fluids, which means a thin wall separating the fluids (while still having the required mechanical sturdiness) made of a material that conducts heat well; a thick wooden barrier would not be good, for example.

The third key consideration is to maximize the area of the wall across which heat is to flow. In addition, we want to have a low distance for heat to flow from wall to center of the inner tube, to speed up internal transfer. This seems contradictory, as we want a big tube (to have a big wall area) but not a big tube (to promote internal heat movement). The answer is to have lots of small tubes instead, giving maximum area without any of them having too large internal distances across which heat must move. The final considerations relate to the flow of liquids, which should help with the mixing of heat within the tubes, as a slow sluggish flow will result in liquid near the inner tube wall heating disproportionately, while a more turbulent mixed-up flow will keep removing the heated liquid and pushing fresh liquid to be heated into the hot zone.

In terms of the flow of liquids, there is also a big difference in efficiency of heat transfer depending on whether, in the horizontal picture I just created, the liquids in the inner and outer tubes flow in the same direction (left to right), or in oppo-site directions (so-called countercurrent flow). If they enter their tubes in opposite directions, the cold liquid encounters gradually hotter liquid as it moves through the system.

A lot of relatively simple yet unavoidably intimidating equations can be avoided here by skipping to the punchline that this type of heating system will result, for cold and hot liquids of identical initial temperatures, in more efficient heat transfer, with the cold liquid hitting a higher final temperature at the exit from the heat ex-changer, all else being equal. In the setup shown in Figure 11.1, the flow is counter-current, and the temperature change between the inlet and the outlet of the inner tube in which the colder liquid flows will be greater (in other words, heating will be more efficient) than if the arrows in the tubes were pointing in the same directions.

The kind of system I have described here is called a shell-and-tube heat ex-changer, with the shell being the outer tube and the inner one being referred to as the tube.

Exactly the same principles apply to heating food in a cooking scenario in a kitchen. Picture heating liquid in a saucepan; the hotter the plate (heat source), the greater the area of the base of the saucepan, the less liquid in the pan (least distance from any of it to the hot base), and the more we agitate the liquid (by stirring) will all help heat faster, while we are unlikely to use a porcelain saucepan as it won't conduct heat well. Similarly, picture a steak on a frying pan. The hotter the pan, the faster it will cook, and thin steaks will cook faster than thick ones, as the heat has less distance to travel to get to the center. In addition, we encounter in this case the principle that excessive heating at the outside of the material is bad not just because it can burn, but because such a burnt layer can form an insulating tough barrier that doesn't easily permit heat through it to the undercooked center (sometimes called case hardening in food heating).

Now, going back to the processing scenario, we can picture a freely flowing liquid rushing through the tubes and shell, perhaps helped on its way by a pump, but what about something much more thick and viscous, like maybe a sauce or ketchup? In such a case, the fluid won't be able to move very easily and won't mix readily, and so the likelihood is it will burn easily on the walls and form an insulating layer, and so the food in the middle of the tube won't heat properly at all. Here, the fluid needs a helping hand, and so heat exchangers used for such products have a rotating shaft down the middle of the inner tube, from the sides of which protrude scraping blades, which continuously scrape the surface of the wall, clearing off the heated material, mixing the fluid, bringing fresh cold material to the wall, and keeping the fluid moving through the tube, all at the same time. Such a heat exchanger is called, in a typical case of literal-minded engineer-speak, a scraped-surface heat exchanger.

In a typical process for heating food to preserve it, there are actually at least three major operations to be undertaken, which are (A) heating to a particular temperature, (B) holding at that temperature for a certain time calculated to make sure the desired effect is achieved (typically killing a particular bacterium, as discussed in Chapter 10), and then (C) cooling back down. For example, pasteurizing milk involves heating from cold storage at 5 °C to 72 °C, holding there for 15–30 seconds, and then cooling back to 5 °C. In the first step, the goal is to get heat into the milk as quickly as possible, while in the third the opposite is the target: energy in, then energy out.

One way to achieve this could involve hot water being used first to supply energy, and cold water later to suck it back out, but both are materials that cost money to produce for food processors. Is there an easier way?

The milk at one stage needs energy, and shortly afterward needs to get rid of it. This clear confluence of interests can be exploited in in a very clever trading-off principle called regeneration, in which the outgoing hot milk surrenders its excess heat to the incoming cold milk, which is heated in the process. In a shell-and-tube system, this can be envisaged as the incoming fluid, which has been heated by flowing though the tube, reversing direction and flowing through the shell on the way back out, after the holding stage, during which it heats the incoming cold liquid that follows.

This makes the process enormously more efficient, and in essence recycles a large proportion of the energy in the system, meaning that a heating–cooling process needs only a fraction of the heating or cooling medium it would otherwise consume. This remaining fraction is still needed because even the best regeneration system cannot completely recycle all the heat so that no additional medium is required, as some energy is always lost from the system, for example, by heating the metal of the heat exchanger and being radiated away from even well-insulated surfaces.

The most efficient heat exchanger of all in essence gets rid of tubes entirely and has all the flow taking place in very narrow gaps between metal plates with corrugated surfaces (to vastly increase their surface area for heat transfer). A plate heat exchanger will have hundreds of plates clamped together in a frame, separated by thin leak-proof rubber gaskets. These plates are arranged such that, in alternating gaps between pairs of such plates, hot and cold liquids are flowing, losing heat or gaining it as required in either direction to the liquid in the neighboring gaps. There are also holes in the corners of the plates that, thanks to careful traffic direction by rubber gaskets, either allow liquid to flow through (and then out of) the gap in plates where it is supposed to be or else pass on and flow through the next gap. Such equipment is used in milk pasteurizers around the world, being highly admired both for their enormous efficiency and the compact size in which a powerful heating and cooling system can be fitted. Similar equipment operating solely with chilled water can also be used on forms to rapidly cool milk after removal from the cows, in which case the systems are called plate coolers.

If one of these plate heat exchangers is responsible for pasteurizing milk, and hence on a daily basis produces milk that is consumed by perhaps hundreds of thousands of people, it is not surprising that such machines have in place very reliable safety systems to prevent the possibility of a huge public health emergency being due to widespread unplanned consumption of unpasteurized or under-pasteurized milk. In one such measure, the milk, once heated to the required temperature (by a combination of regeneration of heat from outgoing pasteurized milk and use of a final heating stage using hot water to bring it the last few degrees) flows through a tube of dimensions precisely calculated to give the required holding time, and then flows past a calibrated temperature sensor. At this point, if the required minimum pasteurization temperature isn't detected, the valve diverts the milk back to the start, while sounding clear alarms for the operators. Otherwise, if the correct temperature is measured, the milk is allowed back through regeneration, cooling, and eventual packaging and dispatch.

Of course, a further potentially dangerous step happens when pasteurized and raw milk are flowing in close proximity to each other during regeneration, separated only by the plates; what if a leak were to occur through a crack or pinhole in a single plate? To avoid any problems should such a failure occur, the milk before regeneration passes through what is called a booster pump, which puts it under pressure such that, should a crack appear in a plate, pasteurized milk could flow only into raw milk (which is at a lower pressure, and so cannot flow in the opposite direction).

In addition, the milk is routinely tested for the efficiency of pasteurization, to further validate that the process has worked (beyond presumably the evidence of a clear lack of stringent consumer complaints and news headlines). How this testing is done provides a useful example of how process validation in general often occurs for food processing.

First, we must consider the actual objective of pasteurization. The original conditions for this process were defined based on the conditions required to inactivate *Mycobacterium tuberculosis,* the bacterium which causes tuberculosis (TB), a disease that, up to midway through the 20th century, resulted in very significant numbers of deaths and cases of long-term hospitalization (and still does today in some parts of the world, while the emergence of antibiotic-resistant strains of the bacterium are of growing concern more widely). This was regarded as the most heat-resistant non-spore-forming bacterium likely to be found in milk, and so if the conditions reached were sufficient to kill it, then all less-sturdy pathogens would be killed also.

To verify whether a pasteurizer had done its job, it might seem logical to test for the presence of this bacterium: no bacteria, job done! However, a negative detection would not be definitive in this case, as raw milk that (as we would hope would be the case today) didn't contain any of that particular bacteria in the first place would test negative, but could still be highly dangerous because of the unmeasured presence of other nasty pathogens. So, testing the target is not the answer and, even if it were, the prospect of testing for a dangerous bacterium, requiring trained personnel and complex media and often taking days before return of a result, would not be practical. Happily, there is an enzyme in milk (alkaline phosphatase) that has many very convenient properties; it is always present in raw milk, it is killed under essentially the same conditions as *Mycobacterium tuberculosis,* and it is easy and quick to test for. So, it plays the role of a surrogate and is routinely tested for pasteurized milk, with a negative phosphatase test being a requirement for release of milk for consumption.

This idea of measuring something easy to substitute for something much more difficult or dangerous to test for is common in food processing and makes ensuring safety in a rapidly moving production system both feasible and thoroughly reliable.

Can we heat solid foods? Yes we can.

Heat exchangers such as those previously described present a range of possibilities for liquids, but it is impossible to persuade solids to flow through a tube, or at least to do so and in any way resemble their original state afterward. How are solid foods like beans, fruit pieces, vegetables, and others preserved for long periods by heat, in cases where their fresh consumption is not always possible?

The answer to this, as we have discussed in Chapter 10 when we met Napoleon and his Euro-conquering ambitions, is to place food in a sealed container, typically of metal, and heat the hell out of it (or, more precisely, to heat the spores out of it!). In such cases, the food and their protective container are immersed in an atmosphere of steam (the source of heat, and a very efficient one at that) and held there for periods typically of the order of minutes, which is much longer than the tens of seconds usually used for heating liquids in heat exchangers.

This time difference is necessary because the main mechanism for heating solids is conduction, which is inherently a slower means of transferring heat than the convection we find in fluids. The solids through which heat must be conducted include the metal walls of the can and the food pieces themselves. We do whatever we can to help the sluggish heat flow, though, by methods such as filling the spaces around the food pieces with fluid, because that will help the heat to flow much better from can walls to food much better, through adding an element of convection far beyond that which would take place if the spaces were filled by insulating air instead. We can even take advantage of this fluid by adding something that adds an antimicrobial hurdle, such as salt or sugar, which makes the environment inside the can that little bit more hostile for microbial survival (adding an additional hurdle). This allows us to apply a less-severe heat treatment, as the bacteria are more sensitive to the heat (hitting a man when he is down may be seen as unfair play in human interactions, but is a key principle in killing bacteria in food). Further, we can help the convection (just as we do when stirring a pot on a hot plate) by mixing the fluid, which in the case of cans being heated means they are rotated or inverted regularly during heating to add an element of mixing.

The simplest piece of equipment used for such a process consists of a metal chamber in which the materials to be heated are placed, a lid is applied and securely closed, and steam introduced or generated (by boiling water in the chamber) to raise the temperature to the required levels. The steam filling the chamber is trapped but highly energetic (the water molecules can be pictured as bouncing off the walls of the chamber, as they seek to find a way out). Sometimes, the air present at the start is vented carefully to allow more room for the steam to fill. This generates very high pressures in the chamber, which prevent the contents therein from boiling (which would negatively affect the quality of the food) because increasing pressure increases the boiling point of water, and so we can readily hold the chamber contents at temperatures of 110–120 °C without boiling. So, one of the main characteristics of such a chamber will be its ability to withstand high pressures (strong walls!) and high temperatures.

The temperature is then maintained for the required period of time to achieve the sterilizing effect required, and the source of heat removed, to begin the cooling process. This can be done by turning off the heat source and/or carefully venting off the steam, or perhaps introducing cooling water. The sudden removal of the energetic steam molecules, or at least their rapid calming from their previously overexcited state, can result in a sudden drop in pressure, which might lead to undesirable physical stresses on whatever is within, such as buckling inward of a container as it cannot mechanically withstand the abrupt change of pressures. To prevent such undesirable outcomes while still getting rapid cooling, air can be gently introduced into the chamber to replace the steam, and pressure gradually decreased, while water being sprayed in also can do this while helping to cool the contents more quickly.

This is the basic principle of operation of pressure cookers, a common feature of domestic kitchens for many decades thanks to their ability to cook rapidly while maintaining food quality, because of the high temperatures involved and short cooking times (as generally any reaction that happens at a certain rate at a certain temperature will happen at a predictably faster rate at a higher temperature). In such cases, the objective is cooking rather than (mainly) killing bacteria, but the principle is the same.

The pressure inside a pressure cooker is generally such that the boiling point is raised by around 20 °C, but a principle called the Arrhenius equation states that most chemical reactions double in the rate at which they take place for every 10-degree increase in temperature, so that being able to increase temperature without boiling greatly speeds up the cooking process (which, after all, is the sum of a whole lot of chemical reactions, such as the Maillard reaction, being triggered in our food). Pressure cookers trap and utilize very powerful forces, and must be used very carefully, or otherwise very serious accidents and injuries can occur.

Beyond the kitchen, this is also the principle of operation of autoclaves, which are used for sterilizing instruments in hospitals and media for bacterial growth and testing in microbiology laboratories.

The equivalent systems to autoclaves (sharing many operational features with these) for larger-scale food processing are called retorts (although not necessarily witty ones). In the food industry, and other scenarios where large amounts of material need to be heated under such conditions, the retorts can get pretty big, and room-sized retorting chambers are possible, into which baskets of food containers are wheeled.

If the objective is to make sure all the food is free of even the most heat-resistant bacterial spore, which if it were to survive could cause illness or spoilage, the goal must be to make sure that *all* the containers and *all* the food in all the containers are heated enough to do this (but no more than this, to prevent both unwanted damage to the food and excess energy wastage).

So, we need to find the part of a batch of food being heated in which this might be most difficult to achieve, and focus our attention in designing the heating process to get the required heating at that point in particular. In a basket of cans, for example, we could picture that a can right in the middle, which is surrounded by other cans that absorb the heat and is also farthest from the source of heat, might be marginally harder to heat than its neighbors, and so we would focus on that can as the target to get right, as achieving the right level of heating there should mean that all other cans receive at least that treatment.

But, considering the coldest can alone is not enough, as there must surely be a point within that can that was harder to heat than the rest of the can contents. If we picture the heat entering the can through all external walls, lid and base, then it must flow through the contents such that eventually all the parts of the food within feel the heat adequately.

If the can contents are completely solid, it can be expected that the last part to which the heat can penetrate is right at the center, as the heat flows by conduction equally from all directions, and every arrow of the direction of heat flow meets in the middle; this is called the "cold point" of this can.

If the can has liquid or partly liquid contents, however, the heat will not travel in straight lines, as conduction is not the only heat-transfer mechanism at work but rather convection also plays a part, and we know that convection involves heated material becoming less dense and rising. So, if a can of liquid is heated without any agitation or mixing, the upper half will heat slightly faster than the lower, as the heat rises. In this case, the cold point will not be found at the dead center (as, in here, the bacteria must die), but below center, because of convection, and, in liquids of different viscosities (as, the more viscous the liquid is, more sluggishly the heat will flow, as convection currents struggle to swirl as they might wish), or mixtures of liquids and solids, the cold point will move between center and below-center appropriately.

A key factor here is doing whatever we can to help the heat transfer within the can. This can be done quite simply by rotating the cans or otherwise agitating them during the heating, so that convection is given a helping hand, and the movement of heat through the can speeded up considerably. Unfortunately, for completely solid foods, agitation has little benefit, but most food retorts will include the capability of rotation as a simple means of helping to reduce the processing time for any food containing at least some liquid, as reducing the time at high temperatures will be a key means of getting the highest-quality food product at the end of the process.

In practice, these conditions are measured and monitored very carefully indeed. This is because, if a process is not operated properly, processors cannot know for certain that their food won't cause illness or worse for their consumer and result in the enormous consequences of a case of something like botulism being associated with their product. To achieve such monitoring, cans can be specially fitted with a temperature probe, carefully positioned at the calculated coldest point of the can, and this can placed at the calculated coldest (hardest to heat) part of the retort.

Running this "probe can" through a full heating cycle then generates a set of data for temperature at that coldest point against time through the process, and this can be used to calculate whether the right degree of safety can be guaranteed.

Going back to the principles of kinetics of bacterial killing introduced in Chapter 10, we can take a dangerous bacterium such as *Clostridium botulinum* and use what we know about its resistance to heat to check our process data. Essentially, this goes back to the fact that we know with high certainty that heating *Clostridium* to 121 °C for three minutes is enough to kill a level of the spores that is far above anything we would reasonably expect to find in food.

So, if our autoclave or retort was able to heat the canned food therein to exactly 121 °C for exactly three minutes, we would be sure we had safety. In reality, though, achieving such control that we can go from room temperature to 121 °C, hold for 3 minutes, and go right back down again, is essentially impossible.[1] In practice, in an autoclave or retort, the product might be held in steam for as much as 30 minutes, but during that time it is slowly heating up toward the temperature of the steam (which might be 121 °C or higher) and mightn't even have reached this by the time the temperature starts to go back down.

Is this not a problem, so, in that our target temperature isn't reached? No, because even reaching 110 °C will kill *Clostridia*, just less effectively than 121 °C, while 115 °C will have intermediate effectiveness. From the D-values, we can derive

---

[1] At least for the contents of a can. The only kind of heating that can achieve such rapid temperature changes is the addition of steam, leading to heating by transfer of latent heat, followed by evaporation to remove the added water and energy in the form of latent heat, leading to incredibly fast cooling; this is the principle of direct UHT processing.

a separate parameter called the z-value, which captures how the effectiveness of killing a particular type of bacterium changes with temperature. This gives us the ability to compare all the temperatures our product might encounter in the retort, in terms of their killing ability, to what we would have achieved at the ideal temperature of 121 °C.

The term that is used to describe this comparative killing ability is the lethal rate, which, when calculated for a particular temperature, expresses how one minute at that temperature compares to one minute at the benchmark temperature of 121 °C in terms of killing *Clostridia* of a specified z-value.

At 121 °C, the lethal rate is equal to 1 (one minute at 121 °C is as good as any other minute at 121 °C!).

However, below 121 °C, the values are less than 1, and below maybe 110 °C the values have so many zeros after the decimal point before we get to an interesting number that we can ignore them, but as the temperature approaches 121 °C the values creep steadily up toward 1.

Then, if we know that the product was in the retort for 15 minutes and the temperature at the cold point increased over that time (reaching a maximum value of 117 °C), we can slice that 15 minutes into one-minute intervals and evaluate just how lethal the heat was at the average temperature for that minute.

So, maybe the first nine minutes had such tiny lethal rates that we can ignore them, but after that we had Lethal Rates for minutes 10, 11, 12, 13, 14, and 15 of 0.05, 0.12, 0.22, 0.34, 0.52, and 0.65, respectively (just as the temperature never hit 121 °C, so the lethal rate stays shy of 1). So, minute 10 was worth 0.05 minutes at 121 °C, minute 11 was worth 0.1 minute and so on. To see how the whole process shaped up as a *Clostridia*-killing onslaught, we simply add these up, which gives 1.9. This number is called the total lethality, and means that our entire heating process was equivalent to 1.9 minutes at 121 °C.

I already noted that the target to get a desirable level of killing of *Clostridia* was 3 minutes at 121 °C, so it could be concluded that this treatment didn't achieve enough heating to make the product safe. It must be remembered, however, that just because we turned off the heat at minute 15 doesn't mean the product cooled immediately, and just as it took quite a while to get to 117 °C, so it will take a while to get back down (probably faster though if we are helping cooling by adding cold water). Still, minute 16 might contribute another 0.55 minutes worth of killing, minute 17 another 0.35, minute 18 another 0.2, and even if subsequent minutes don't add much to our sum we are now up to the magic value of 3 minutes of equivalent *Clostridia* carnage for our process, and we could rest assured that the product should be safe.

On the one hand, if our full calculation, including heating and cooling, led to a value of 2.5, we would conclude that we weren't heating enough and would have to either increase our temperature of steam or the time in the retort, or both. We would change the system accordingly, rerun the temperature measurement with the can fitted with the sensor, and see if we are closer to the target (mathematical predictions can help to make this less trial and error than it might sound).

On the other hand, if our calculation gave a value of 3.8, we could conclude we are overshooting and overcooking, going beyond the degree of heating needed to achieve our safety goal and probably damaging the food in the process, and we would amend time, temperature, or both, accordingly.

By this approach, we can predict what conditions are needed to ensure consumer safety and check that a process delivers exactly this objective while minimizing unnecessary damage to the food.

We can also apply exactly the same principles to understand what the effect of times and temperatures on quality attributes like what the content of a vitamin might be and understand exactly the conditions we need to avoid if possible to prevent such changes. Following from this, we can also use the same principles to model the exact conditions needed to achieve a desirable change while avoiding an undesirable one.

As mentioned in Chapter 10, we can also adapt the method for heating processes other than the severe conditions of canning by using a different reference temperature for comparison than 121 °C (which historically came from concern about *Clostridia*), such as 100 °C for boiling vegetables or 72 °C for pasteurizing milk.

Through these tools, plots can be drawn that show a huge range of temperatures and times encountered in food processing, like a map of all possible combinations of both, but crisscrossed by lines that show the exact conditions that will result in specific effects, like an acceptable level of kill of *Clostridium* or *Salmonella*, or the destruction of 90% of vitamin C, or the development of an off-flavor or odor. Processors can then pick a set of conditions carefully that makes sure the good stuff happens while avoiding the bad stuff.

Going back to the practical considerations of retorting food in essentially a large rotating pressure cooker, this is an example of what we referred to already as a batch process. To initiate a heating cycle, a tray or basket of cans or other containers are loaded into the machine, the heating starts, time passes, the cooling starts, the cans (now sterile) are removed, and a new batch is loaded. Start, stop. Load, run, unload. High labor, much downtime, and a high proportion of each day spent not actually heating but doing the steps in between.

Batch processes are very much unloved by the food industry for these reasons, but for many years a significant technological challenge lay in how to make a retort into a continuously operating system. To understand this, picture a conveyor belt loaded with cans passing through a chamber filled with steam, within which they spent long enough to be rendered sterile, after which they left at the far side. During this operation, cans are continuously entering the chamber at one end and leaving at the other, as is needed for it to be a continuous process. In this case, though, the problem is the fact that the active agent with which the cans come into contact is steam under pressure. To achieve pressures of this sort requires entrapment, or otherwise the steam will find a way to escape, leading to loss of pressure and likely great injury to those nearby; such a degree of entrapment is hard to achieve when there must be two points of opening, in other words, that through which the cans enter the chamber and that through which they leave. Solving this problem was sufficiently hard that batch retorts remained common in most canning operations (and still do in some places) for many years.

Cracking the steam-chamber problem was achieved by exploiting another principle of fundamental physics, which is that a tall column of water exerts a pressure at its bottom that is proportional to its height (the term for this is hydrostatic pressure, reflecting that fact that it is caused by stationary water). In a continuous can-sterilization retort, tall columns of water flank a chamber of steam, in a shape that could be pictured as a letter W, with the sides being vertical and the inverted V in the middle being a steam zone. If a conveyor belt is then fed into the top of the left-hand leg or column, loaded with cans, it descends downward and is exposed to progressively higher pressures (but nice and slowly), and then enters the middle bit where it is surrounded with steam, and heats accordingly. The speed of the belt is controlled such that the cans each spend exactly the right amount of time in this central steam heating zone to achieve the required heating intensity (checked with an occasional can fitted with an internal temperature logger), and then they leave, entering the right-hand leg or column, and slowly ascend, pressure gradually decreasing as they climb. The belt exiting the column can then be sprayed with water or pass through a cold-water bath to help cool the cans, which then get to a point at which they are removed from the belt, to be replaced by fresh cold cans to be heated, and the process continues, continuously.

How does this solve the problem? The pressure of the water columns traps the steam in the middle without leaking out and maintains sufficient pressure that the contents of the cans don't boil during the processing. This technology is what is likely to be found in any large-scale canning operation today.

Cooking in the kitchen

In the previous section, we considered the type of heating that is applied to food in an industrial scenario, whether a pasteurizer or canning system. Of course, on a day-to-day basis, we apply exactly the same principles when we cook food in our kitchens.

For example, when we use a grill we apply radiant heat (radiation) to transmit energy across the air gap onto the surface of the food being grilled (assuming we define grill in the European sense, which is the from-above cooking method sometimes called broiling in the United States). The more energy we apply by turning up the heat on the grill, the more energy is pumped onto the food surface and the more rapidly its temperature will change as a result. We know that if we don't get this right we can burn our steak, toast, or whatever.

The heat in this case is localized at the surface though, and below the surface there will be cooler material so, thanks to the irresistible demands of thermodynamics, heat will flow from hot to cold and start to penetrate downward through the food. Assuming the material being grilled is a solid, this will be by conduction, and so the heat will travel in straight lines toward the center, and then onward as fresh unheated regions of the food become accessible to the march of the heat from the surface. If we left the food being heated from one direction, the heat would gradually penetrate through to the far side, and this would heat from the inside outward, being shadowed from the main external heat source by the rest of the food. To heat a piece of meat in this way, such that the far side eventually cooked, would be rather unwise, though, as to get the heat through to have the far side, say, cooked medium rare, would result in a huge trail of damage as more and more heat was applied to the surface exposed to the heat source, which would as a result be somewhat on the incinerated side of well done.

To avoid this, we flip the meat regularly, to expose both sides to the direct effects of the radiant heat, and so initiate heat transfer from both sides equally (it can even be shown that regular flipping heats more quickly than a single flip halfway through cooking). Then, the last part to heat isn't the far side of the food, but rather the center, and the key question then becomes heating in such an even way at the right rate so that both sides plus center have a consistency (plus a level of safety, don't forget the importance of safety) that is acceptable and desirable.

Several common heating techniques used in the kitchen will similarly apply heat unevenly, and rely on the chef to account for this by regular turning or mixing of the food to expose fresh surfaces and directions to the heat. When we fry, for example, heat comes from the hot plate through the metal base of the pan by conduction, usually with the assistance of a thin, convective layer of oily liquid, to one surface of the food, so we flip regularly to ensure that both sides get cooked and become a source of heat heading for the core.[2] The oil also helps to fill in gaps and uneven patches on the food surface, to make sure heat is transferred evenly.

We also make sure that we optimize the conduction effect by making our pan from a material that conducts well and so transfers the heat it receives from the hot plate below evenly into the food being fried. Copper is very good for this purpose, even better than stainless steel, and so gives more even heating across the pan's surface, but is liable to wear and break down, especially if exposed to acidic conditions, while iron pans are highly sensitive to rust. Aluminum, by comparison, retains heat well (heating up and slowing down more slowly than other metals), but is also acid-intolerant. Some frying pans might combine layers of metals to get the best heat-transfer properties (such as pans with a copper layer sandwiched between other metals[3]), but overall stainless-steel pans are probably most common because of the balance of advantages and disadvantages mentioned.

The thickness of the base of the pan is also critical, as a thicker pan, irrespective of the material, presents more of a barrier to heating, and so, on the one hand, will be slower to transfer the heat, but, on the other hand, will transfer the heat more evenly and cool more slowly when the heat source is removed. For these kinds of reasons, different people, even for different applications, might prefer different

---

[2] We do the same when we heat from below using flames from coal or gas (again, a source of direct radiant heat) on a barbecue.

[3] This is analogous to how packaging is often constructed, as will be discussed in Chapter 14.

kinds of pan, and the key factor is the way in which the pan takes the heat to which it is subjected and transmits it to the food therein.

The inner surface of any frying pan today, however, will rarely find metal in direct contact with food, without some intervening protective (for the metal) layer. This is the principle of the nonstick frying pan, which has been treated in such a way as to prevent the kinds of reaction between food and metal that would otherwise result in burning or sticking and, later, stressful cleaning.

As discussed in Chapter 3, when we heat food to very high temperatures the proteins unfold (denature) and can stick to each other and form complex structures, but can also equally form bonds with metal surfaces. After running something like milk through a system like a pasteurizer or UHT plant for prolonged periods of time, the metal interior surfaces can be seen to gradually accumulate a white coating (called "fouling"), sometimes composed of precipitated minerals that have been made insoluble by the applied heat and sometimes composed of proteins. Throughout this book, we have seen examples of where heat causes food molecules to change their properties and interact with other molecules, and also we have seen how this can be key to desirable changes during cooking. However, it is perhaps surprising to envisage the same kinds of reaction taking place between food and an inert metal surface, giving a weird food–pan hybrid (even for those who might feel they need more iron in their diet), but this is exactly what can happen in the absence of some kind of barrier that keeps metal and protein apart.

A further consideration will be avoidance of the food staying in contact with the metal surface for long enough for such reactions to take place, so we should keep meat moving around the pan while heating or the contents otherwise well stirred.

In practice, most modern nonstick pans are coated with a layer of what is called poly(tetrafluoroethylene), PTFE for short, or Teflon for less-scary acronym-ism! Teflon is completely inert to the food and, of course, so as not to interfere with the cooking process, is a good conductor of heat, and the applied layer is so thin as to present very little obstacle to the transfer of heat for cooking.[4]

In the case of an oven, we rely generally on convection-driven upward air currents in a sealed (mostly, except when we open the doors) chamber to transfer heat from the heating elements (and the hot walls in particular) to the food. We know already from the nature of how convection heats the air in the oven that the upper regions of the oven will heat faster than the lower ones, as the hottest air rises. We can even that out and avoid such differences by using a fan in the oven to artificially speed up convection, giving more controlled and rapid cooking. We also want as little barrier in the way between food and heat as possible, and so place the food on rack shelves rather than solid ones, so that containers can be held in place but heat can flow easily through all the gaps between the thin metal bars of the shelves.

---

[4] I am mindful here of the joke about how the Teflon layer is made to stick to the nonstick pan, but let's just say "well done chemistry" (in all possible ways of reading those words) and leave it at that.

The food then is typically at least partially protected and supported by a metal container like a tray, pan, or dish, which will heat by convection in the oven and then transfer this heat to the food surface with which it is in contact by conduction, and again start the movement of heat through the cooler layers of the food by conduction (when solid) or convection (when it is liquid) or some combination of the two (for example, when cooking a casserole containing solid pieces of meat or vegetables in a sauce or broth). The rest of the food surfaces not in contact with the tray then receive their heat from the hot air within the oven, so further heat flows in different directions are initiated.

A key consideration is whether the food will be covered during this coking step or not. If moisture loss is likely to be a major factor, and considering that evaporation of moisture will oppose our efforts to cook the food by absorbing energy in the form of the latent heat required to enable the transformation of water from liquid to vapor, we will cover our food with a lid or foil, for example. In addition, a lid will entrap the water vapor and fill the space above the food with highly energetic steam molecules, which in their agitated but futile attempts to escape increase the pressure inside the pot, and so depress the boiling point of the water therein, allowing more-intense heating to occur at lower temperatures.

Many of the same principles will then apply if we replace air as the source of heat, as in an oven, with water, as when boiling in a pan. The surrounding medium in this case transfers heat into all surfaces of the food simultaneously, and thus a gradient is established in which heat travels from every direction toward the core at once. Keep a lid on the pot and the pressure builds, encouraging boiling and cooking, while removing the lid reduces the pressure and can calm the turbid waters within. Boiling in such a situation actually begins in microscopic cracks and imperfections across the bottom of the pot, as steam begins to gather here and inflates into bubbles, and, as the water heats to boiling point these bubbles break free, and colder water gets drawn into the gaps left empty, to repeat the process. The bubbles are less dense than the liquid and rise as a result, reaching the surface and bursting, which cools the surface (as latent heat is absorbed), unless the pressure is too high for this to happen, as would happen if there were a lid on the pot.

Overall, different heating methods applied in the kitchen cook food in different ways because they apply different heat-transfer methods to get heat into the food. Sometimes we rely on radiation to pour heat onto food from a hot surface, as when we grill, bake, barbecue over hot charcoal, or (as, we will see next) use a microwave. In other methods, as when we steam, boil, sauté, or deep fry, we are relying on conduction and convection to transfer heat.

Searing a piece of meat relies almost entirely on conduction, while convection can play a role in heating systems that are either "dry" (such as roasting or deep-fat frying[5]) or "wet," as when we boil, poach, or steam, for example. This

---

[5] Here, although there is liquid involved, it is oil rather than water, and so we consider it as dry heating. We can hit much higher temperatures with hot oil than hot water, though, which is why

difference in the key means of transferring heat leads to differences in rates of cooking, as grilling will transfer heat more rapidly than an oven, for example. Final temperatures reached can also differ between methods, as wet methods don't typically reach as high temperatures (because of the cooling effects of water evaporation) as dry methods, and so the cooking-related changes we associate with very high temperatures such as Maillard reactions and caramelization won't occur to the same extent.

When we deep-fry in oil, we can achieve very high temperatures, as the hot oil can heat much higher than water could because of evaporation. Of course, the food being deep-fried contains water, though, that will evaporate, at first from the surface, when that food is placed in the hot oil. This generates steam bubbles that escape from the food and heat upward for the surface (because of their lower density than that of the oil), and this has the advantage of mixing the oil and drawing hot oil to the surface while preventing too much oil seeping into the food. When the surface is still moist, evaporation will keep it cool, but when it dries the temperature will start to climb rapidly.

In addition, when we stop supplying heat, as when we take something out of the oven or turn off the grill or hot plate, we don't just turn off heat transfer, as heat will continue to move from hotter regions (like the surface) to cooler ones (like the center) by the simple rules of thermodynamics; this is called "carryover."

Sous vide cooking: Not-quite-boil in the bag cooking

This method of cooking was introduced in France in the 1970s and involves heating food in a water bath (a large metal bath in which water temperature can be precisely controlled) at exactly our target cooking temperature. This gives very exact control on which reactions off the menu of possible heat-induced changes in food we want to allow to happen, and it is possible to ensure that such reactions occur evenly throughout an entire piece of food.

Before heating, the food is placed in a plastic bag from which the air is removed under vacuum (*sous vide* in French), and the bag is sealed. The bag keeps the heating medium (the water) away from the food, and so everything in the food stays in the food, and flavors or nutrients, for example, are not lost in the cooking water, as happens in conventional in-pot cooking. Also, compared with conventional cooking, the temperature doesn't fluctuate or need to be modified by turning up or down the cooker controls, as the mechanics of the water bath keep the temperature at exactly the set temperature.

The very precise temperature control could be used, for example, to cook an egg such that the relative balance between degrees of coagulation of white and yellow

---

deep-fried food can show much deeper color (and sometimes flavor) development than if the same food were boiled in water.

parts can be exactly specified; the egg mightn't even need to be packaged, coming as it does in its own natural package.

In theory, the surrounding source of heat doesn't have to be water, and could be oil (to get to much higher temperatures), or even air, but air is a much weaker heating agent than water, so is little used.

One key consideration in *sous vide* is that temperatures used are not too low that bacteria that might cause food poisoning are not killed; having a very unusual egg texture is no good if salmonellosis results. Ultimately, the key advantage is being able to precisely control the extent of cooking applied, so that every piece of a product can be cooked to exactly the same degree, offering a level of consistency that cannot be achieved with most cooking methods. The elimination of oxygen can also be an advantage for foods such as fish or fruit and vegetables, where its presence could lead to problems that are due to oxidation or enzymatic discoloration, and of course its absence helps to discourage the survival of the high proportion of bacteria that require oxygen for their survival.

Heating food from the inside out: The microwave

What if we could somehow, as if by magic, generate heat in food by exciting the water molecules themselves without application of an external heat source? Hot water is basically the same as cold water (once we don't change it into steam or ice); the molecules are just moving faster and, the hotter the water, the faster the molecules are moving. To make water hot, we can either supply the energy externally to do this or perhaps could imagine somehow grabbing each molecule and spinning it rapidly, like a finger flicking a child's spinning top.

How could this happen? The secret to such water-twiddling was discovered several decades ago and still forms the basis of operation of the microwave ovens found in many kitchens today. The basic principle was actually modified from technology developed during the Second World War to detect enemy planes by radar. It is hard now to reconstruct the mental leap from something that filled the sky to something that could be captured in a small box, but a gentleman called Percy Spencer took just that leap (inspired by a chocolate bar melting in his pocket when he passed by a microwave generator) and bridged the gap from warfare to homeware in a few short years, in a modern version of turning swords into ploughshares.

The modern microwave oven heats the food within it by passing microwave radiation through it. This is a form of energy that obeys the same principles as more familiar forms of radiation such as radio waves and light but is of a different physical form. The field microwaves generate when passing through water in the oven chamber essentially creates forces at exactly the same scale as water molecules. When microwave rays encounter water molecules, as they pass back and forth through the chamber, they buffet them like waves passing a small boat, causing them to spin, and neighboring molecules to thereby bump and rub into each other, which generates the energy we feel as heat.

The heat is generated specifically in the water-rich parts of the food, which explains why it is typically advised to leave food stand after removal from the oven, to allow the heat to travel through the other regions, by conduction through solid parts, and to even out the cooking. This time is referred to as the relaxation time (perhaps because the chef can take a breather while waiting for it to happen?) and the length required for it to happen can differ between food materials.

The inside-out heating also explains why food doesn't brown or change color in the same way as food heated in a traditional oven, where the heating happens from the outside-in, and the surface reaches the highest temperatures, and the reactions that develop color can take place there. In the microwave, temperatures hot enough to give the browning effects such as the Maillard reaction are simply not reached.

One of the other peculiar aspects of microwave cooking is that the air surrounding the food is not hot, unlike conventional ovens, and so the ability of the air to drive or encourage dehydration from the surface is much less, and so products do not dry as they might in other types of cooking and can even appear comparatively soggy after microwaving.

How cool is that? Low-temperature food preservation

For a very long time, people have known that keeping food cool slowed down the rate at which it perished. Storing food in ice, cold streams, peat bogs, or even just cool underground cellars has been practiced around the world where conditions permit since prehistory.

Everything moves more slowly in the cold, whether people shivering on a frosty morning or bacteria far from the balmier temperatures at which they most like to grow. As we have seen repeatedly through this book, a high proportion of changes in food, particularly the less-desirable ones, are biological phenomena, which mostly have been optimized by millennia of evolution to perform best at body temperature (around 37 °C). Reactions, whether driven by chemistry, enzymes, or microbial growth, largely require a level of energy that is optimal around this temperature, and slow down or stop the further we stray from this ideal.

To take bacterial growth, being in many ways the primary concern for the spoilage and safety of our food, most bacteria are what are called mesophiles, growing best at that value of 37 °C. As mentioned in Chapter 7, some like it hot, though,[6] and we have already discussed at length the problems that thermoduric bacteria and their spore-armor can cause food processors and, if not eliminated, possibly the public. Here, though, we go in the other direction on the temperature scale and focus on the psychrophilic bacteria, who like it cool.

There are several families and species of bacteria for whom the temperatures of around 5 °C found in a typical fridge are perfectly hospitable, and these are

---

[6] Well, nobody's perfect!

the main problem we might encounter in a refrigerated product. Thankfully, most, but not all (as we will see), such bacteria are more of concern for reasons of food spoilage than food safety.

In milk still warm from the cow, certain types of bacteria will dominate if it is left standing uncooled, but once we (ideally rapidly) refrigerate it, we shift the balance to conditions most favorable for psychrophiles, of which the most common in the farm environment are called *Pseudomonas*. These bacteria will multiply rapidly in the cold but are not generally a safety risk. They do, however, have a particular significance for milk processing in that they produce enzymes that can break down milk proteins and fats, often with destabilizing or flavor-ruining effect. While the parent bacteria are so cold-adapted that exposure to even mild heat (less severe than pasteurization) will wipe them out, the enzymes they produce and secrete into the milk are tough as nails and can survive even the extreme temperatures applied during UHT processing. So, control of *Pseudomonas* growth is a major priority for this reason (particularly through not leaving milk unheated for so long that they get a chance to grow to levels where these enzymes become a problem for the later quality of dairy products).

In terms of nonspoilage bacteria that grow at refrigeration temperatures, the most notable is probably *Listeria*, because of its pathogenic nature[7] and in particular its significance for certain cohorts of individuals, such as pregnant women and immunocompromised people.

It is not just bacteria that can work in the cold and cause problems during refrigeration, though. Consider the metabolism of fish, which has been optimized to work at cold water temperatures (close to those found in a refrigerator). When we catch fish and store it in the cold, these reactions continue to take place, because, compared with warm-blooded meat, this doesn't actually represent a sudden change to unaccustomed circumstances, and such reactions can lead to the distinctive odors we associate with fish.

For the majority of food products, though, the key is to get the temperature away from the troublesome temperatures around that of the body at which enzymes and (most) bacteria are most likely to wreak their damage, and into a chilly region where the slowdown in such activities means we can keep our food safe and unspoiled for longer. How can we best achieve this?

Let's start by considering the domestic refrigerator found in most kitchens around the world, or in larger scale in stores, factories, etc.

In this case, food is cooled in what is essentially a box from which we need to continuously pump heat to the outside. No matter how well we insulate this box, there will be a temperature difference between it and the outside world, which tends to drive heat into the box, while we regularly assist this by opening the door (allowing warm air in) and placing food inside that is perhaps not already at the same temperature as the refrigerator. So, all this heat needs to be continuously scooped up from

---

[7] Should a pathogen that grows in the fridge and can cause death be called a psychropath?

the inside and dumped on the outside, and we need a sort of circuit by which something absorbs heat inside the refrigerator and releases it outside, and then repeats the process, in an endless cycle.

One of the processes in physics that is best at absorbing heat is a change in state, such as evaporation. Many materials, including water, become extremely energy-hungry when making the leap from liquid to gas or vapor, if encouraged to attempt this change in their condition. We have seen in Chapter 9 that the easiest way to make materials change more easily from liquid to gas or vice versa is by manipulating pressure. So, picture a liquid that would like to be a gas, but is kept under pressures that force it into the less-exciting fluid condition; allow it to expand by allowing it to fill a larger volume, and thereby exposing it to a much lower pressure, and it will gleefully make the leap to gas, sucking in energy from its environment to fuel this escape.

This is the reaction that takes place out of sight at the back inside wall of your fridge, and the energy the material swallows is that which needs to be taken out to keep the fridge at the target low temperature.

The gas then leaves the chamber part of the circuit and is put under pressure again, restricting its movement and forcing it to condense back into a liquid. In this process, it must eject the excess energy it now has, and this is jettisoned as efficiently as possible by allowing the liquid to flow through a narrow set of metal pipes with a very high surface area, which we can see easily on the outside back of a fridge and feel, if we place a hand nearby, as the escaping heat.

The energy-depleted liquid then completes its external circuit and passes back into the interior part of the system, where it is allowed to expand, turns to vapor, and by absorbing the energy to do so cools the interior, and the food therein.

What a cool cycle that is! We just need to be careful not to overstress the system, as when we add in a sudden injection of heat that is transferred to the cold contents of the fridge, warming them up in a potentially dangerous way before the pump can get on top of things and evacuate the excess heat. This is what we do when we put something that is too hot into the fridge.

In this way, we apply thermodynamics in our kitchens every day, and exploit physics to influence biology, and keep our perishable food for longer without replenishment or spoilage. To take but one example, a product like pasteurized milk would be infeasible as a safe commodity without the power of refrigeration to keep it safe, as we have seen.

Then of course we can go further, and indeed deeper on the temperature scale, and use our iceboxes and freezers to store our food at subzero temperatures, at which we don't just slow down many of the reactions of concern but essentially turn them off entirely.

The principles of heat removal may be similar to those in our refrigeration system, or we might use a scraped-surface heat exchanger to freeze our ice cream mix just before consumption, or aim for the least-damaging type of freezing we can manage by using ultra-cold high-speed air in a blast freezer or fluidized bed. In all cases, we need to take advantage of the many preservative benefits of freezing while avoiding the damage that expanding the volume of a key component of our food in an uncontrolled manner can cause major damage (as discussed in Chapter 9).

Then, as in refrigeration, having brought our food to the desired low temperature, the key is to keep it at a steady temperature. Opening a freezer door and allowing a tiny amount of warm air in doesn't raise the temperature enough to encourage microbial growth to kick off, but it can be enough to induce a small amount of ice to convert to water. When the door shuts and the freezer brings the temperature back down, this ice will refreeze, but as discussed in Chapter 6 for ice cream generally results in ice crystals getting larger, and larger means more damage to the food structure. So, even if we store something for months in a freezer, if that time period included a few even tiny increases in temperature, the cumulative effect will be to induce a progressive increase in jagged crystals forming, to ultimate disadvantage to our food.

So, temperature, be it high or low, is a powerful tool in the hands of the food processor, and I include anyone in a kitchen within that grouping, but in both cases has to be managed, controlled, and optimized. Otherwise, the benefits for our food can be outweighed by the detriment caused, whether it be an steak well done to the point of carbonization or a piece of meat dredged from the depths of a freezer that has become essentially a solid block of ice, most of which has escaped thanks to crystal damage and accumulated on the surface like a severe hoary frost.

12

For My Next Trick

MAKING WATER DISAPPEAR

The benefits of drying

Compared with some of the processes we have discussed so far, like heating or cooling, drying is one we might think less of in a kitchen context and consider to be more a large-scale industrial process.

However, when we look around a kitchen we find a lot of products of such activity, in terms of containers of powders like salt, sugar, spices, milk powder, soups, flavorings, flour, and much more, as illustrated in Figure 12.1. These have enormous advantages of long life, not needing to be kept in the fridge, taking up relatively little space, and providing a neat and concentrated source of whatever flavor or other character we wish to add to a dish. The key consideration is that whichever powder we use will behave in a convenient way when we come to use it, dissolving in water or other meal bases easily and reliably.

We also routinely remove water from food in the kitchen, perhaps not by having a mini–spray dryer on the counter (at least not in most kitchens), but by removing a lid from a pot to allow some water to be driven off in the form of steam. We also essentially remove water from food more subtly, for example, by adding sugar to a jam recipe, which does not remove the water as such, but rather renders it less available for undesirable things like supporting microbial growth, thereby achieving many of the stabilizing and preservative benefits of actual drying without the drying.

Removing water from food greatly improves the stability of food products and, by inhibiting the actions of microorganisms, increases its safety. As a result, drying of food, wholly or partially, has been practiced for centuries as a way to make food more stable.

Not only does removing water from food add major hurdles in terms of stability and safety, but it adds an enormous bonus feature of convenience. To illustrate this clearly, I always think of what life would be like if we had to buy all our coffee in liquid form, and no dried (or highly concentrated pod) versions existed. Think how many gallons we would be dragging home from supermarkets or stores every week, how much storage space would be needed to keep it (probably chilled), and how we would need to heat it up cup by cup when we needed it. Thanks to drying, on the other hand, we instead buy a tiny percentage of the weight of liquid coffee as a stable dry powder or concentrate that we can keep for weeks or months on the shelf and add hot water to when we need it.

Drying food to a powder, however, is another example of a deceptively complex operation, as the hard part isn't necessarily getting the water out, but rather relates to the fact that what will ultimately be consumed is not the powder, but a mixture of the powder and water. For this reason, the ability of a powder to reconstitute (dissolve, to all intents and purposes) in water readily (ideally with minimal effort, which is the defining property of what we call an "instant" powder) is probably its most important property and, for some food products, one of the most difficult to get right.

One of the oldest industrial processes for drying food from a liquid state into a powder is called roller-drying, which involves pouring a thin layer of the liquid to be dried onto the hot surface of a rotating metal drum or roller. While the liquid lies on the drum, heat from the hot surface drives evaporation of water molecules from it, and the solids in the liquid are left behind, to be scraped off by an appropriately positioned blade as flakes. The newly denuded drum surface is then coated in fresh liquid, and the process repeated. This is simple and reasonably effective, but is a pretty harsh process, which gives a rather overheated dry powder and is rarely used today for this reason.

Of course, solids can be dried as well as liquids and, for centuries, solid food has been dried by simply leaving it exposed to air, ideally hot air. It is believed that ovens were originally used more for drying than for cooking food. To dry food, we intuitively do some of the things that scientific rationales indicate will accelerate the rate of drying, such as maximizing surface area, maximizing the exposure of the food to the drying medium, and having that drying medium (air) be as dry and hot as possible. So, today, we can have cabinets in which trays of food are exposed to hot air, or systems in which the process is both speeded up and made more gentle by applying a vacuum, so that drying takes place at lower temperatures (in vacuum desiccators).

The drying of liquids

In terms of drying liquids, roller-drying was gradually supplanted from the 1950s onward by spray-drying as the technology of choice for drying products such as milk.

When I teach the principles of food processing, I always tell students that the process that will surprise them most if and when they see it in industrial practice is spray-drying. I can discuss and explain it in lectures and say in my most earnest tone that these are really really big and impressive bits of equipment, but until one has seen for real the scale and scope of these tremendous beasts of food processing, they cannot truly be appreciated.

When driving past a food processing plant, it is easy to guess if they have a spray-dryer, as this will be obvious if there is a very tall building sticking up into the sky, perhaps 10 or more stories in height.

Inside that building, there will be found a structure that consists of a very large cylinder, the bottom of which tapers into a cone, with perhaps some rectangular boxes clustered around the base. Hanging onto the side of the main cylinder are a few smaller versions of the large chamber, and all sections are joined by a network of pipes. The inside of the building will be hot, as heat is radiated off the dryer, and often exceptionally noisy, thanks to large hammers that at preprogrammed intervals clang off the walls of the chamber to dislodge any powder that may have adhered there. In addition, one of the walls of the building will likely be somewhat different to the rest, as it is designed to blow outward (into an unoccupied part of the plant site!) should the enormous forces being put to industrious use in the dryer get out of control and cause an explosion. Finally, the main drying chambers often have small windows or hatches to permit inspection of the process within, and anyone peering through such an opening will see a snowstorm-like blizzard of powder tumbling ever downward (see Figure 12.2).

What is going on in this megalithic machine?

Let's first start with a liquid to be dried, for example, skimmed milk (from which fat has been removed) being dried into skim milk powder or coffee being dried. Before going anywhere near the dryer, some of the water in this liquid will likely have been removed in advance, to make the drying process more efficient.

This typically is done in an evaporator, in the main parts of which a film of milk flows down the inside of multiple metal tubes[1] that are surrounded by a heating medium such as steam. As the liquid flows, some of the water evaporates at the high temperatures encountered, and vapor rises, while the progressively more concentrated and heavier milk sinks. In this way, vapor (water) can be removed from milk, which becomes more concentrated as a result. In a modern milk evaporator, the pressure in the evaporation chamber is reduced by drawing a vacuum, so that evaporation can take place at lower temperatures than would normally apply, as discussed earlier, resulting in less heat damage to the milk and a better final product. In fact, if the process is split over a number of different connected stages (called effects), at a progressively stronger vacuum (and hence lower boiling point), the vapor from one effect can be used as the heating medium for the next effect in the series, as it will be at a higher temperature, and so the process can be made extremely efficient in terms of reducing the amount of steam required and increasing the reuse of energy in the system.

---

[1] The thin films and multiple tubes directly exploit the principle encountered in previous chapters of maximizing the area over which a desired process takes place.

Anyway, by a method like this, the skim milk has perhaps around half the target amount of water removed by evaporation and is also now hot, and so is at this point fed into the top of the spray-drying chamber.

To dry in this chamber, we aim to have the liquid in the form in which it exposes the absolute maximum surface area for the volume to be dried, which is a spray of droplets.

Such a spray is produced by a process called atomization, of which there are two common types. The first is very similar in its operation to a garden hose and uses a nozzle to create a final steady spray of droplets. The second uses a disk-shaped bowl with lots of holes or slots around its outer edge, into the center of which the liquid to be atomized is fed as the disk rotates at high speeds. Under such circumstances, the liquid is thrown out through the holes or slots as a fine mist of droplets.

The droplets produced by either atomization method then encounter a circulating stream of exceptionally hot (typically over 200 °C) dry air, which has been fed into the chamber in a swirling rotary flow pattern, as shown in Figure 12.2. Each individual droplet then falls downward under the irresistible force of gravity, surrounded and pursued by this hot stream of air, which first strips the water from the outer surface of each droplet, then sucks moisture from deeper and deeper within the drop as it falls. Toward the bottom of the chamber, the air and former droplets, now converted into dried powder particles, part company as they leave the chamber. This is achieved thanks to carefully channeled flow patterns that funnel air and powder through different exits, giving in theory one stream of dry powder and one stream of air, the latter now significantly cooler and moister than when it entered the chamber.[2]

In practice, though, a number of modifications are usually required or added to this basic process. The first is necessitated by the fact that the separation of dry powder from drying air at the end of the process is driven by the weight of the powder particles themselves, and a significant fraction of these particles (called fines) will be too light to neatly separate, and rather swoop out of the chamber with the outlet air. This air thus enters the smaller cylindroconical[3] structures mentioned earlier, called cyclones, where a clever arrangement of geometry and air flow give a powerful separation force, which strips the fines from the air and recovers them for adding to the powder leaving the main drying chamber.[4] To be sure, the air can pass through several cyclones to increase its eventual purity,[5] or perhaps in some systems

---

[2] The temperature of the outlet air is actually a key indicator of exactly how much drying went on in the chamber as, the more drying that took place, the cooler it will have become because of the energy it has surrendered to the liquid as it turned to powder, and this temperature can be used to fine-tune the drying process itself.

[3] Part cylinder (the top bit) and part cone (the bottom bit).

[4] The same separation principle can be found more domestically in many modern vacuum cleaners (like the Dyson vacuum).

[5] Which is important for reasons both of process efficiency and environmental impact.

can be passed through a porous layer of material that allows air through but traps powder, called a bag filter.

So, the powder is captured at several points (main chamber plus one or more cyclones—only one is shown in Figure 12.2 for simplicity—and possibly a bag filter) and then blended and packaged in suitable bags. This is called single-stage drying and is the (relatively!) simplest and cheapest type of spray-drying. The powder thus produced is not completely free of water[6] but, if we have only around 2%–4% water left, we regard it as dry enough to be stable, as such a low level of water will discourage microbial growth and most chemical reactions over weeks or months of storage. Also, if the material was much drier, it would actually become very hard to prevent it from absorbing moisture from its surroundings, often to undesirable effect.

However, single-stage drying has a number of disadvantages from the key perspective of the quality of the finished product and what it can be used for. Obviously, the most efficient cheap drying process in the world is no good if the product it produces cannot be used for anything, and in the case of powders the main property that determines their usefulness is their ability to be dissolved in water, as mentioned earlier.

The reaction of powder and water, and the instantization of powder

When a spoon- or dispenser-scoopful of powder, whether it be soup, coffee, infant formula, milk, or any other dried material, is added to water, a series of highly complex phenomena take place.

The first concerns how the material disperses in the water on first contact. Does it clump into lumps, break up easily, or float on the surface? Factors such as the presence of water-repelling materials will immediately make the first reaction of water and powder an awkward one. For example, milk powder containing fat (whole milk powder) is much more difficult to dissolve than skim (fat-free) milk powder thanks to the uncooperative hydrophobicity of fat, and such powder typically needs to have an amphiphilic[7] agent added that can bridge the water and fat materials peacefully to overcome this problem.[8] In contrast, the presence of water-loving substances such as sugars draws the water into the powder, having the

---

[6] This is because some water will be too tightly bound to food constituents to be removed under conditions that are reasonable to use in a feasible process.

[7] As previously described, this refers to a molecule that has part of its structure that is hydrophobic and can react well with fat, while another region is hydrophilic and reacts well with water. Such a molecule is happiest at a water–oil interface, where it smugly gets the best of both worlds.

[8] Such powders are referred to as being lecithinated, as lecithin (an amphiphilic phospholipid found in many food materials such as eggs and soya beans) is the agent added, and the principle of action is as discussed in Chapter 5 for emulsifiers in terms of the bridging action between hydrophobic and hydrophilic food components.

opposite effect to the presence of fat. A further important factor is the water temperature as, the hotter the water, the easier the dispersion,[9] while the presence of air spaces in the powder creates pathways and tunnels for water to get into, allowing it to dissolve the powder from the inside out. Finally, as always, a bit of physical forces helps the process too, be it shaking a bottle with infant formula plus water or whisking a bowl of powder and water.

Then, once the water enters the powder, the particles must come apart totally, with the water-soluble elements (such as sugars) dissolving in the water happily, while those components that are not exactly water-soluble (like proteins or fat droplets) become dispersed evenly enough in the water that they remain in suspension without settling out, at least for as long as we need them to.

Powders that achieve this state easily (least time, least need for temperature or tiring physical exertion) are referred to as instant, and in general this is the desired goal of any food-drying process.

Unfortunately, the spray-drying process, in which all the water is removed in a single stage, is not always very successful in producing instant powders, and so more advanced processes were developed that optimized the structure of the powder particles produced, and hence their reaction with water and ultimate usefulness and value.

Such processes can even tailor the size and shape of the particles and how much air they contain, as air is a key enabler of instant behavior. It is known what shape and range of sizes of particles facilitate the best incorporation of air both within (called occluded air) and between (called interstitial air) particles. The likelihood of entrapping air within an individual droplet, to expand toward the end of drying to give a nice water-welcoming cavity at the heart of a particle, can even be controlled.

Consider two samples of powder, one containing a lot of trapped air and one containing far less; the former has the disadvantage of taking up a lot more space in a bag or tin, but this must be traded off against the gusto with which it will imbibe water when the consumer needs it to.

To tailor these dissolving properties requires significant modification to the spray-drying process, particularly in terms of not requiring all the drying to take place in the main chamber, but spreading it out over a few stages, across which temperature and air flow can be individually and sensitively controlled.

One way to do this is to allow the powder to exit the chamber at a higher moisture content, maybe 10%, and then enter a fluidized bed, as introduced in Chapter 9 for freezing vegetables, but in this case being a tunnel through which warm, dry air carries the powder particles falling from the main chamber; these are the rectangular boxes often found below the main chamber. In the fluidized bed, the still somewhat damp powder particles float along under circumstances that positively encourage them to bump into each other, stick together, and eventually dry to the

---

[9] We can picture hotter water as containing more energetic molecules of water, which bash into the powder and smash it up with more and more vigor as the temperature rises.

target final moisture content as larger so-called agglomerated[10] particles, which entrap the air that is so useful when water must later be reintroduced to the product. This can be seen at the bottom of Figure 12.2.

Images of different kinds of milk powders taken with an electron microscope are shown in Figure 12.3, and the strange structures that can be produced are seen. Some powders have a wide range of sizes of particles, as seen in the middle image, which is important as these can pack very densely in a package, with small ones nestling in between larger ones in a way that is much more efficient than could be achieved if they were all the same size. Some particles might be large but hollow, with big air spaces within, as seen in the shattered particle in the top image, while the bottom picture shows the complex structures that can be conjured into existence in a so-called agglomeration of powders, to optimize their ability to dissolve readily in water when needed.

Such drying processes are called two-stage drying, as the drying takes place first in the main chamber and then in the fluidized bed. In some dryers, the bottom cone of the main dryer is truncated, and the particles fall directly into an integrated circular fluidized bed for some initial agglomeration before entering the attached external fluidized beds; such systems are called three-stage dryers.

So, in this way, the drying of products such as milk can be carefully controlled with the goal of optimizing the properties of the final product that are most important to the end user. Further processing of the powder, such as grinding the sugar shown in Figure 12.4, can change the properties further, with particle size and surface area (increased with smaller particles for the same weight of powder) being known to greatly increase the perception of saltiness or sweetness in the mouth, for example.

For some products, as we shall see, even the most careful spray-drying is not ideal, in particular because of the presence, in even the most carefully controlled system, of very hot air.

Freeze-drying and the quest for ever-better coffee

One product for which this consideration has driven the development of alternative drying processes to spray-drying is coffee. The main disadvantage of spray-drying for coffee is the fact that the hot air, when escaping from the dry product, strips off any molecules that are easily encouraged to enter a gas phase rather than staying behind in the dry solid powder. If we picture molecules that readily become gas and

---

[10] To picture what these look like, it is useful to consider the technical terms that are actually used to describe their structure when visualized under a microscope, which include "raspberries" and "grapes." The complex structures of milk powder particles can be seen in Figure 12.3, where a range of sizes, shapes, and degrees of clustering (all deliberately created by control of the drying process) can be seen, as well as the large air bubbles that can be created within individual particles, the better to imbibe water when being dissolved.

float into air, we can imagine that these are critical to one of the most important properties of a product such as good coffee, which is aroma; the aroma of coffee comes specifically from the capture of such flavorsome escapees by the bank of sensors that is in our nose, and aroma is a far more valuable attribute of a product like coffee than it is for, for example, milk.

To dry products while avoiding such inadvertent and unwanted stripping of desirable aromas, an alternative process is often used, called freeze-drying. This is another example of a food processing technique that was apparently practiced long before the science that makes it actually work was understood, as legend has it that the ancient Incas of Peru effectively practiced this by drying food at high altitudes on cold mountainsides, where the warming of the morning sun induced what was basically the same effect as that of highly complex machinery now used to achieve high-value coffee powder.

Freeze-drying is also another example of a process that exploits some of the weird properties of water, in this case the ability to make it, under certain rarified conditions, do something it normally never does in common experience. Some materials on heating can skip the liquid state entirely, going on heating from a solid to a gas; this is called sublimation, and the trick can genuinely be regarded as sublime (but on the right side of ridiculous).

To induce water to adopt this behavior, which means that ice can transform to vapor without bothering to exist as liquid in between, requires placing ice (or a frozen food sample) under conditions of very low temperature and very low pressure (vacuum).[11] Under such conditions, application of very gentle heating[12] to the product causes the water to vaporize and leave the product behind. This happens because the freezing point is not affected by decreasing pressure in the same way that the boiling point is, so at a sufficiently low pressure the boiling point equals the freezing point, and the only transition below this pressure that is possible is that from ice to vapor (liquid water cannot exist).

Now, the key principle in drying is obviously to remove water from food, and so this is a form of drying but, compared with spray-drying, the huge difference is the very low temperatures used. As discussed earlier, the problem with drying at high temperatures is the stripping off of valuable volatile (gaseous) aroma compounds, but the low temperatures used in freeze-drying don't encourage the volatiles to follow the water, and so they stay behind, leaving a much richer and complex flavor for the dry product.

In addition, when the product was initially frozen, we know that crystals of water were formed throughout it and, on freeze-drying, these essentially disappear and leave behind in their place air cavities, which we have already agreed are the key to something that will reconstitute quickly when water is added. Thus,

---

[11] The link to the mountainsides is that, at higher elevations, the pressure is lower, as mentioned in previous chapters, and so elevated slopes supply the low pressure part of the requirements for freeze-drying (which is very Ande[s]).

[12] For example, by the first rays of light as dawn breaks on a Peruvian mountainside.

freeze-drying gives a material that is highly porous and will respond very favorably when reintroduced to (liquid) water.

Looking at different types of coffee down a microscope shows this very clearly. The upper image in Figure 12.5 shows a cut surface of a freeze-dried coffee granule, and large bubble-like pores are everywhere, while spray-drying gives the much denser and less porous structure seen in the lower image.

In practice, to achieve freeze-drying (assuming the absence for practical reasons of a convenient mountainous environment), a frozen product is placed on trays or in suitable vessels in a chamber within which a vacuum is drawn, after which the trays or vessels are very gently warmed; water sublimes off and is removed from the air within the chamber by exploiting its tendency to condense as ice on the surfaces that are providing the cooling effect. If a tray of liquid is dried, the resulting product has the volume and appearance of the original material but can readily be seen to be a ghostly light structure that only resembles what we started with, and can easily be broken up into a light, fluffy powder. If a solid material such as a steak or a piece of fruit were freeze-dried, the resulting product would look like the starting material, but would actually be composed largely of air.

A process such as this has certain unavoidable disadvantages as a processing technique; the scale of the systems is relatively small (compared with the enormity of spray-dryers), the process must be operated batch by batch (as opposed to continuously), and it is very slow (drying the product in hours or days, compared with minutes for spray-drying). All of these factors make freeze-drying a very expensive process, and so its use is justified only for very high-value products that can justify the additional processing expense (because alternative types of drying just don't deliver the characteristics that consumers expect); high-quality flavorsome coffee is a perfect example, as are fruit pieces and spices.

How else can we control water in food?

So far, we have considered removing water from food at high or low temperatures, and seen how different product considerations influence the selection of process used. There are also a number of other options that may be used, however, particularly if our goal is not necessarily to remove almost all of the water.

For example, we know that when we freeze food, we convert some of the water to ice, which leaves behind a concentrated remainder of the food. This can be exploited to partially freeze a liquid like fruit juice, stock, or broth, after which the ice crystals that are produced can then be physically removed by filtering them out, leaving behind a concentrated liquid.

At the other end of the temperature scale, we have seen already how evaporation can be used to combine high temperatures and low pressures to concentrate liquids by removing some of the water; this can be used not only as a precursor to spray-drying but as a means of concentrating products such as evaporated milk to be used for baking, by giving a source of milk solids and flavor without having to add a lot of water as well. Evaporated milk has been produced since the 19th century for this purpose, where milk was concentrated, placed in cans, and sterilized in a canning retort, to give a concentrated source of milk flavor and compounds to add in baking applications, for example.

In addition, a close relative that has been a feature of kitchens and bakeries for centuries to evaporated (or condensed[13]) milk is called sweetened condensed milk. In this case, the product is given a very long life not by heating the can to very high temperatures, but by exploiting a different principle of how we can remove water, or at least make it less available to support undesirable phenomena such as microbial growth. This is achieved by adding to the evaporated milk a very high level of sugar (sucrose), which effectively binds up the free water. Any bacterial cell finding itself in that environment will rapidly die through dehydration as osmotic pressure draws water out of the cell in an ultimately doomed effort to equalize sugar levels inside and outside the cell.

In the kitchen, we routinely remove water from food by boiling off water from a pan on a hot plate, to thicken or concentrate a soup, stew, or stock, or prepare a reduction, although such methods can drive off many flavorful volatile aroma compounds. To reduce heat-induced damage during this step, we can use lower temperatures to achieve similar degrees of evaporation when we draw a vacuum (because we know that lowering pressure reduces the boiling point of water), and more adventurous modernist kitchens might include a machine in which the liquid to be concentrated is spun in a rounded flask in a water-bath, at reduced pressures, while the water that evaporates off is condensed and collected in a long neck protruding from the flask.

We can also use extremely fine filter materials called membranes to filter food products, such that a liquid is separated into fractions that can (permeate) and cannot (retentate) pass through incredibly fine pores in a suitable material. In the process that is commercially used that has the very smallest pores (reverse osmosis), only water molecules can pass through the membrane, and so water can be removed with very fine selectivity and temperature control, to concentrate very sensitive materials, or indeed to recover very pure water! Such processes are discussed in more detail in the next chapter.

---

[13] Condensed milk is a common name for what is actually evaporated milk, and is a somewhat ironic name, as condensing is the opposite of evaporation, and the milk did not condense, but presumably the water that was removed from it eventually did.

13

Squashing, Filtering, and Other Ways
We Process Food

The need for alternatives to heat

As I have discussed throughout this book, mankind has relied on heating to preserve and make safe our food for a very long time, even long before the science of how and why this works was understood.

However, clearly using heat to process food is a rather blunt tool, sometimes as subtle in its effects as hitting it with a club or bat. Just as it kills bacteria, molds, and other microorganisms, it inflicts collateral damage on the sensory and nutritional quality of the food. The greater the level of kill, and hence stability and safety conferred, the greater damage to the "fresh-like" characteristics of the food has usually been caused.

A question could then be posed as to whether, instead of applying such a crude and damaging (although undoubtedly effective) treatment, we could treat food with more of a surgical-scalpel or laser-focused treatment, which zoomed in on and very specifically destroyed the target microorganisms while leaving the surrounding food as little changed as possible.

This is the target of so-called minimal (sometimes called invisible) processing, and today there are a range of technologies that have promise for achieving this goal. Indeed, because of the desirability of such an outcome, this has been thus one of the most active areas of research on food processing in recent years.

High-pressure processing

We have encountered the importance of pressure several times already in this book, usually in how its manipulation can affect properties of water such as boiling. Pressure has another important application in food processing, though, in that it can replace heat as the physical force we apply to achieve desirable change in food.

In food processing circles, high-pressure (HP) processing is often referred to as a novel processing technology, but in fact it has been around for quite a long time. Remarkably, around the same time that Pasteur was explaining how heat works in terms of preserving food, on the other side of the Atlantic an American scientist called Bert Hite at the West Virginia Agriculture Station was doing experiments on his own homemade pressure-generating apparatus. From this work, Hite showed that foods such as meat, juices, and milk could be preserved by the bactericidal effects of HP.

Legend has it that he became ever more curious about this new technology and continued testing samples of ever-more-dangerous cultures, until there was a sudden and presumably unwelcome eruption during one experiment, spewing dangerous material about the lab. Apparently, and really not that surprisingly, his colleagues persuaded him to cease his endeavors at this point, and Hite, at the height of his pioneering work, stopped his work under such pressure, which was no doubt high.

Around 1910, another American, Percy Bridgman, showed that an egg, when subjected to high pressures without the application of heat, coagulated; this was the first demonstration that pressure, as well as killing bacteria, could denature proteins, and thus have other potential applications in food processing.

There, however, the historical trail goes rather cold, presumably because of the lack of availability of suitable processing equipment, and these prescient and pioneering pressurization processes remained undeveloped for most of the 20th century. In the meantime, though, lots of other fields of engineering became interested in pressure, and systems were designed and built for nonfood applications such as materials, ceramics, synthetic diamonds, and aircraft parts, whereby giant chambers in which such materials could be fused or shaped with incredible precision at gigantic temperatures and pressures became quite widely used.

In this light, it is not surprising that eventually the circle would close again, and the food applications of such equipment, without the need for the ultra-high temperatures, would be rediscovered. This led to the emergence of a range of HP-treated food products, such as rice and purees, in Japan around 1990. Since then, HP treatment has been on a rapidly expending trajectory as a food processing technology, finding applications in particular where existing technologies struggle to deliver a high-quality consumer-ready product.

It is accepted that HP treatment will remain a relatively niche technology, however, as it remains limited by the scale of available equipment (small by standards of the modern food industry) and cost (the opposite!). For these reasons, HP is finding its place for (microbial) "killer" applications where HP offers something that heat or other processes simply cannot deliver.

To take one example, let us consider the oyster. In recent years, there has been a lot of talk about convenience foods, but oysters represent an inconvenience food, for many reasons. For one, they come in their own natural defensive package, or shell, which is difficult to open without damaging either meat within or opener without. For another, they live in aquatic environments where they simply absorb any passing materials, including bacteria and viruses, leading to occasional roles as vectors of rather inconvenient food poisoning.[1] To render them safe they may be heated, but then their flavor and texture become quite different and less desirable to those who value such slimy slithery salty stuff.

Around 20 years ago, an American company called Motivatit, who were major oyster harvesters in Louisiana, heard about HP processing and, on learning that a local oil-drilling company had such equipment, decided to put a bucket of oysters under pressure to see what would happen. What happened was rather unexpected; the most dramatic and obvious effect was that the shell was released,[2] becoming opened or "shucked," while the meat within became somewhat juicier, through apparently absorbing the small amount of liquid typically found in the shell. Microbiological testing indicated a positive impact on safety and shelf life as well, and so HP processing had achieved what no other processing process could deliver: a safer, more convenient oyster. A great marketing angle was then developed whereby the oyster was surrounded before processing with a yellow band that held the shell closed during HP treatment. The band could then be snipped when desired for easy release of the contents, and Gold Band Oysters became an enormous success.

To actually achieve HP treatment, a food product to be treated is normally placed in a suitable flexible package (often the final package in which it will be sold) and loaded into a cylindrical chamber (with very thick walls to confer enormous strength) that is then filled with a pressure-transmitting medium, such as water or an emulsion of oil and water. This medium is then compressed by very powerful pumps or a piston directly pressing on it, and the pressure thus generated is transmitted uniformly and rapidly to the food within. The food is subjected to pressures from every direction at once, and so it should not be imagined that the force applied is a crushing or deforming force, once air is not present to collapse within the structure of the food. Treatment times are typically of the order of a couple of minutes (some minutes are also required for the buildup and release of pressure), following which the product is removed and the cycle repeated.

An important question is how high is high pressure? We saw in Chapter 8 that HPs are found on earth below the surface of the earth or sea, and we can readily picture in *National Geographic* documentaries or movies such as *Titanic* the thick walls of submarines or submersibles that are used to probe the deepest parts of the ocean. From such images, we have perhaps a sense of how hazardous such journeys are for those who undertake them, and how surprising it is when (frequently bizarre) life-forms are found on such trips. At the deepest parts of the ocean, such as the Mariana Trench, pressures are around 1000 times that found at the surface.[3]

---

[1] An Irish tradition of combining consumption of oysters with that of Guinness is not, despite the preservative effects of alcohol, regarded as a suitable food safety strategy.

[2] Because of pressure-induced denaturation of proteins in the muscle (as opposed to mussel) that holds the shell tightly shut.

[3] The units used in HP processing are megapascals (MPa), and the pressure at sea level is one tenth of one of these units, or 0.1 MPa. The pressures at the deepest points on the ocean floor (like the Mariana Trench in the Pacific Ocean) are then around 100 MPa.

Within the chambers in which oysters and other food products are treated, the pressure is at least around 2.5 times this pressure, and often up to 6 times as high.[4] So, the conditions applied exceed those that are found anywhere on earth where humans or their robotic extensions have ventured. In fact, such pressures would be found only on tunneling deep deep underground, heading for the earth's core.

On this basis, it is perhaps not surprising that such conditions are remarkably hazardous for undesirable microorganisms in food, such as the bacteria and viruses in oysters, which are used to much milder pressure conditions. It can also affect molecules in food, though, as Bridgman showed when he heat-lessly cooked his eggs a century ago, and in some cases its preservative interest comes about because of its impact on other agents which cause undesirable quality changes in food, such as enzymes.

For example, another early success for HP processing was in preservation of the avocado puree that makes guacamole, which is highly perishable because of the activity of enzymes that cause color and textural deterioration rapidly after manufacture. In this case, HP treatment can inactivate these enzymes and maintain product quality for far longer than unprocessed guacamole, whereas application of heat would have completely changed the character of the product.[5]

So, for oysters and guacamole, HP processing offered a solution to problems that were previously regarded as intractable. For other products, additional interest in use of HP arises from the key difference between how it exerts its effects and how heat affects food; this relates to the types of bonds that are affected by the two processes.

On the one hand, heat affects a type of bond within molecules called covalent bonds, which are pretty fundamental links between atoms in molecules, whether large or small. Pressure, on the other hand, does not affect such bonds, and focuses its disruptive energies on rearranging the structures and functions of large molecules, like proteins, for example. The key effect of pressure comes about through the agency of water, which seeks to minimize its volume when subjected to very high pressures, and essentially shoves parts of molecules out of its way, with frequently permanently structure-altering results. So, if a protein is a critical part of a bacteria's metabolism, pressure can kill the bacteria by messing with said protein or, if it is a critical enzyme, that can likewise become fatally inactivated.

Small molecules, though, with limited complex structures, are essentially immune to the effects of pressure and sail through such conditions unaltered. It just happens that many compounds most responsible for desirable sensory and nutritional properties of food, such as vitamins and minerals, as well as color and flavor compounds, fall into this category. For these reasons, HP treatment has the potential to deliver what might be called the Holy Grail of food processing, which is

---

[4] In other words, 250–600 MPa.
[5] Another pioneering American company, Avomex, developed this process for guacamole processing by HP.

microbiological inactivation without accompanying damage to those characteristics of food that most confer freshness (so-called minimal processing).

The food product type in which this "best of both worlds" effectively has been most clearly demonstrated is fruit juice. If we consider freshly squeezed orange juice, this has high sensory and nutritional quality, but is unstable because of enzymes and perhaps acid-tolerant bacteria and yeasts and molds growing in it.[6] To control these agents of unpleasantness traditionally required heat treatment, but heated juice unquestionably loses some of its fresh character. HP treatment, on the other hand, should give the shelf life through microbial and enzymatic inactivation while retaining the full sensory and nutritional profile. For this reason, many successful applications of HP around the world today concern pressure-treated (sometimes called cold-pressed) fruit juices, smoothies, or blended fruit–vegetable juices.

Revolting food: The use of electric fields

Electricity, despite its intrinsic importance in almost all aspects of our daily lives, is obviously a force of danger if not handled correctly, but electric fields, on a small scale, can also be applied to preserve food. This involves essentially electrocuting bacterial cells and thereby increasing the safety and stability of food, again without the application of heat.

In a food processing technology based on this principle, called pulsed-electric-field (PEF) processing, a food product, for example a liquid, is exposed using suitable electrodes to an electric field that keeps rapidly switching the direction of electron flow. This simply confuses the metabolic balance of cells placed therein, eventually leading to holes being punched through the membranes and walls that surround the cells, leakage of their contents, and ultimately death. Every living system depends for life on principles like those of electricity, with nerves and brain impulses being driven by phenomena that can readily be disrupted by surges of electricity (as can be seen in cases of death from electrocution), so it is perhaps not surprising that bacterial cells cannot withstand such treatment and can readily be killed.

This has been commercially applied to the processing of fruit juices, for almost the same reasons as HP processing has found an application there, in other words, stabilization without loss of apparent freshness. Good results have also been reported for whole fruits and vegetables, milk, soup, eggs, and even meat. Thus, PEF treatment can achieve for certain food systems the key goal of minimal processing to achieve stability without sacrificing quality. Beyond this, the ability of a PEF to unlock the pores in cells' walls, as it were, can allow valuable components to be extracted, such as increasing the yield of juice extracted from fruit or sugar

---

[6] The high acidity (low pH) is a major benefit, as it discourages the growth of spores, which pressure alas cannot kill. This fact means that preservation of juice is much simpler than that of a less-acidic product in which spore survival would be much more of a problem.

extracted from sugar beets. For this reason, applications beyond just killing bacteria and other microorganisms have been proposed, with the only caveat being that the technology doesn't inactivate enzymes like heat would, and so products need to be kept under refrigeration to keep these potentially negative influences under control.

Distillation and the separation of water and alcohol

As discussed in Chapter 8, one of the physical properties of water that is exploited in different ways in food processing and cooking is evaporation. However, when water isn't the only liquid present, things become both more complicated and more interesting, particularly when the second liquid present has boiling tendencies different to those of water.

The key example of this concerns mixtures of water and alcohol, such as arise from fermentation as discussed in Chapter 7, which can be fractionated based on the fact that alcohol will boil at lower temperatures than water. This has been exploited for centuries to give a whole range of distilled spirits, from whisky to rum and vodka. To achieve the high alcohol levels associated with such products would be impossible by fermentation alone, as the powerful preservative effect of alcohol would inhibit the fermenting microorganisms into inactivity long before the level got high enough.[7] This typically happens around 20% alcohol.

So, to get higher levels of alcohol than this, we thank the yeast for their service and dismiss them, turning instead to physical means. The key means of doing this, if we can't make more alcohol be present, is to instead make less water be present, as this is the key diluting factor.

Making concentrated spirits starts with the fermentation of some suitable sugar source. This source might be fruit, grains, potatoes, rice, or coconut. Many different sugar sources will do the trick here, as they can all produce alcohol after fermentation, but the final product characteristics will depend to a huge extent on which was chosen and the other compounds and flavors they contribute besides the alcohol. Vodka comes from potatoes, wheat, or other grains, rum from sugarcane juice, brandy from grapes, whiskey from malted barley, and tequila comes from the extract of a cactus-like plant called the agave (which may explain some of its more hostile effects).

Whatever the starting material, the fermentation is allowed to proceed until a level of 5%–12% alcohol is reached, and the mixture is then heated in a closed pot or a still, to start the process of distillation. When boiling begins, a mixture of water, alcohol, and volatile flavor compounds (in other words, those that easily convert into a gaseous state) will rise from the surface, shake off their liquidic shackles,

---

[7] It is hard to avoid the image of the yeasts in the case of alcoholic fermentation passing out in a sedentary stupor because of excessive levels of the fruit of their fermentative labor.

and reach for the sky (or at least the air above the surface of the boiling liquid); this mixture is called an azeotrope and cannot be separated further by distillation, as the components are in a stable balance or equilibrium with each other.

This gas phase then rises in the still, and would escape happily unless it met something that provided a tempting reason for it to stay or change state into a form that can easily be recovered. In this case, the key is to provide cool surfaces, which cause the gaseous mixture to condense, just as a cool mirror in a warm steamy bathroom becomes a magnet for water droplets that create a fog on its surface.

So, the dilute alcohol solution from fermentation is heated to convert all or most of the alcohol, and some of the water, plus the interesting volatile flavor compounds from the fermentation, into a gas. This is then separated from the liquid by simple virtue of its tendency (being less dense and far more mobile than the liquid) to rise, until it is captured by condensing it, giving a new liquid phase that is much more concentrated in alcohol and flavor and represents the distilled spirit.

During the process of distillation, the earliest and latest fractions to come off the fermentate, or so-called "head" and "tail" fractions, can contain some undesirable compounds and may be discarded, while the bit in between is where the rich desirable spirit is usually found.

After a single step of distillation, however, the alcohol content reached might not be sufficiently high (maybe 20%–30% alcohol), and so the process is typically repeated, usually two or three times in total, to reach the target alcohol level, which may be 50%–90%. The same effect may be, in modern large distilleries, achieved more easily by using long columns filled with cooled plates, through which the gases from heating the fermentate rise, and in which the distillate becomes progressively purer. The fermentation mix is often actually heated at the top, with liquid flowing downward by gravity, while the gases eventually rise steadily upward and are collected on the plates arranged throughout the column.

This is a nice example of a continuous process, where raw material is continuously fed in and products removed, while a traditional still (which continues to be used in some processes for very fine whiskeys and brandies, as it is believed to allow greater control of the final product) is a batch, start–stop, process.

Immediately after distillation, the spirit will not taste or even look anything like the desired final product, and needs to undergo, as with so many fermented products, a prolonged period of controlled storage to transform it into the desirable final valuable drink sold to the customer. For some more basic spirits, like vodka or gin, the key step may simply be adding water and flavor to get the target flavor profile and alcohol content, whereas for spirits like brandy and whiskey the product is aged for periods from months to years to develop the desirable flavor and color through a complex set of chemical reactions between the compounds originally present. Traditionally, this is done in wood barrels, where the wood is far from an inert material providing protection and holding but is an active participant in the reactions that take place within, as molecules leach out of the wood and contribute directly to the flavor and color of the spirit within. Based on this key role of the wood, it is not surprising that different spirits are specifically aged in casks made from specific types of wood, which lend unique character to the drinks in question.

Filtering and separation of food

Separating solids from liquids on the basis of their size is a very familiar operation in every kitchen, like when we recover pasta or other particulates in a sieve, the holes in which trap that which we want to retain while letting the water in which the cooking has taken place flow through.

From basic school science classes, we might also picture a familiar experiment where we take a mixture of solids, such as sand and water, and pour it through a folded cone of paper in a funnel; as in the sieve, the water pours through, while the solids are retained. In this case, the paper used looks whole, without visible holes, but close examination under a microscope would show that it is in fact quite porous, with holes of the right size to let the water through while trapping the sand. These paper-based filters can be seen more commonly for filtration of coffee grounds in coffee shops or home systems.

In food processing, filters are used that make these holes look like vast, gaping pores. To put the separation power of these filtration systems in context, let's look at what is being separated. We have already considered the sieve (large food particulates) and the filter funnel (small but visible particulates). What we consider now goes far beyond the visible to the realm of the utterly invisible. In a typical coarse sieve we might have pores of the order of a few millimeters, while a fine kitchen strainer (e.g., for flour) might have pores a little less than a millimeter across and, in the filter paper or coffee filter, the pores are around a hundred times smaller, at 10–11 thousandths of a millimeter, also called micrometers or microns.

Now, let's consider filtering a liquid food like milk, which would flow straight through our sieve or even that filter paper. In a process called microfiltration, however, the pores are of such a size that bacterial cells and large (but still invisible to the naked eye) structures in food like the fat globules in milk that we met in Chapter 4 are retained by the pores, while the rest of milk components (like proteins and lactose) flow through with the water. This can be used to increase the safety and shelf life of milk without (or with less) heat, and gives a product called Extended Shelf-Life Milk that is found in some countries as an alternative to pasteurized or UHT milk. In this case, though, the things being separated are still (relatively) big, like a bacterial cell or a fat droplet. The size of pores required for such separations is around 0.1 micron, or a hundredth the size of those in our paper, and 10,000 times smaller than those in the strainer.

Another process, called ultrafiltration, works on a different scale entirely and has such small pores that protein molecules are actually unable to pass through; these molecules are barely visible with even the most powerful electron microscopes, but are rejected by the pores of ultrafiltration membranes. Ultrafiltration pores are around 0.01 micron in size, or a thousandth the size of those in our filter paper and

100,000 times smaller than those in the strainer. These membranes are commonly used to "fish" the valuable whey protein ingredients out of huge volumes of whey produced in a cheese plant and purify them away from the much smaller molecules of sugars, salts, and water that make up most of that whey.

Going even further into the molecular selectivity spectrum brings us to something called reverse osmosis. We met osmosis earlier in this book as the helpfully preservative process by which water flows across a cell membrane, pulled by differences in salt or sugar concentration on either side of that barrier. In the case of reverse osmosis, we are making water move as well, but in this case by forcing it through such infinitesimally small pores that one of the smallest molecules known, just two hydrogen atoms stuck to one of oxygen, is the only thing that can squeeze through. This can be used to concentrate fluid streams or even to remove salt from water and thereby make it more potable. It is a little hard to measure exactly the size of pores in these membranes, but they can be estimated at around 0.0001 micron, or tens of millions of times smaller than those in the filter paper. To put these figures in comparison, a bacterial cell is typically 0.2 microns in size, while an elephant is 2 meters long; the proportional difference in size that can be achieved between these types of filtration materials (paper to reverse osmosis membrane) is about the same as the ratio of size difference between a bacterial cell and an elephant. That is some serious precision in putting tiny holes in material!

For any of the three processes mentioned, microfiltration, ultrafiltration, and reverse osmosis, the material used to filter is not made of metal or paper, because these couldn't be machined with the degree of precision and control needed, and in the case of paper just wouldn't have the strength for large-scale process use. In the case of microfiltration, with the biggest pores of the three, ceramic membranes can be used. They wouldn't, however, be formed into a conical funnel like the paper, but rather organized in tubes through which the material to be filtered is pumped, with the wall of the tunnel being perforated with the separation pores, such that the water plus whatever else can pass through the pores flows through these (and is referred to as permeate), to be captured in an outer tube. The remaining liquid inside contains whatever cannot pass through (retentate, as it is retained), with which it becomes progressively more concentrated.

For the other two processes, ceramic filters can't even give the degree of precise separation needed and so polymer (plastic) membranes are produced that have, in an incredibly thin layer on their surface, the separating layer with its molecular-level sieving ability, supported by a thicker layer of porous plastic support through which the liquid that has permeated flows toward the point at which it is collected. These polymer membranes may be arranged in tubes, or spirals, or sandwich-like plates, whichever allows the maximum amount of surface for separation to be exposed to the liquid.

The finer the pores in the membrane, the harder it is to induce liquid to flow through them, as might be expected, and the greater the pressure needs to be applied to the force liquid to squeeze through the gaps. In such cases, though, it can be imagined that some of the molecules or larger components that can't fit through the pores get stuck there, wedged nearby or even partly blocking a pore, and so long runs of these systems tend to get curtailed by such "fouling," and creative solutions to remove such debris on the filters have been developed to help the process operate for as long as possible in as efficient a manner as possible.

Overall, these membrane filtration systems are found today in plants from cheese whey processing to water purification and even juice processing, and are a long way from a humble kitchen sieve, although once again the principles are exactly the same. From pasta to proteins, sometimes we need to grab something out of a watery surrounding and, at whichever scale we need this to happen, a solution has been found.

Three-dimensional printing of food

A well-established process in food processing is called extrusion. Within a machine called an extruder, food raw materials, typically rich in starch and protein, are mixed and worked by one or two auger-like (picture a corkscrew) screws under pressure and usually also at high temperatures, being gradually pushed toward the end of a barrel as a homogeneous mass of hot, pressurized material. At the end of this barrel, they pass through a narrow opening, the shape and dimensions of which are carefully controlled, into open air. This results in a rapid reduction in pressure and temperature, which causes the extruded material to expand and set quickly because of rapid cooling, giving a light and expanded solid food structure. This has been used for many years to produce products from breakfast cereals to pasta to crispy snack foods—all products with airy, porous structures. Creative designs of the die at the end of the barrel that forms the gateway between inside and outside allow a high degree of design of the extruded product's shape and texture.

In very recent years, there has been an upsurge in interest in what is essentially the next stage in evolution of the process of extrusion, whereby more fine control of the extruded material can be achieved and much more complex structures generated as a result.

This is three-dimensional (3D) food printing, which is probably one of the most high-profile developments in food processing in recent years, whether at kitchen or factory level, but interestingly finding probably far more interest at the former scale.

In 3D printing, a suitable material is extruded, usually at lower pressures and temperatures than in the older classical extrusion process, through an orifice or die onto a surface on which it adheres and may possibly set, for example, because of a sudden reduction in temperature on hitting a colder surface. If the die were immobile, a pile of material would eventually build up, not particularly appealingly. However, in 3D printing the die or nozzle is mobile, and its mobility is controlled by a computer program directing, through a series of motors and gears, the position of the die or print head in three dimensions on the surface, which can be envisaged as left, right, and up/down.

So, for example, the movements of the die could trace a square that is filled in with printed material, and then the position of the die is raised a little, and a second layer is completed, and so forth until a cube is built; this type of 3D printing is sometimes referred to as additive manufacture or layered deposition, which reflects nicely this sequential building of layers. Finally, the program that defines the shape, and the attached mechanical drivers, can also control the rate of extrusion of the material being printed, by controlling the rate of feeding in of the printing material or the rate at which the end of the plunger on a syringe-like dispenser is depressed to push out the material.

In a basic plastic printer, the raw material is a thin filament of plastic, fed from a coil to the print head, at which point a brief application of high temperature melts the plastic and makes it flowable; the movement of the print head then controls where the plastic comes out, and contact with the cooler printing surface causes rapid solidification of the plastic and setting it in place. Creative programming of the "shape file" in the controlling computer can lead to the printing of an almost infinite variety of shapes and objects, limited only by the size of the printing space available, the amount of plastic raw material, and the user's imagination.

Today, everything from replacement body parts to machine parts and decorations are routinely 3D printed, and printers have even been run in space, so applications of 3D printing are expanding every day, with reports of objects as large as cars and houses being printed.

The question then is how this can be applied to food, which is after all not the same as plastic. In some of the first applications described for food, chocolate was printed into a wide range of imaginative shapes and products, relying on the ability of chocolate to melt and solidify at very defined temperatures and flow reasonably well (if rather messily) at temperatures in between.

There has also been wide success in broader applications for printing confectionery, whereby sugary mixes can flow and then set into all kinds of intricate shapes or be solidified from a bed of powder, giving detail and precision far beyond that which most pastry chefs could manage with a piping bag and the steadiest of hands, for example, in the design of a birthday cake with a child's face or name in 3D glory on the top.

Going beyond these initial high-profile successes is a huge area of interest, from chefs wishing to use such printers in their kitchens (and a few restaurants marketing their uniqueness in this regard already) to food companies thinking about how, in the future, we might have such printers in our kitchens and need to buy raw materials and instructions for how to experiment with our own concoctions. We might imagine buying "pods" of materials to put into our printers, which might look like a high-end home coffee machine.

A key principle of the 3D printing "movement" since its emergence has been that the printers are inexpensive (being bought in a kit in a box a bit like a regular computer printer, with some home assembly required and batteries not included), and accessible to even the least engineering-minded, with methods, recipes, and files for printing shapes being widely shared for download online. So, 3D printing is deliberately not a complex and inaccessible technology, intimidating for the faint-hearted, but rather one with egalitarianism designed in from the start, as it were.

Going back to the idea of the "pods," what might be in them? In recent work in my own laboratory, a group of students showed that certain types of processed cheese can be readily melted and printed using a syringe-like assembly (with the printer moving the nozzle of the syringe to the position specified by a computer file describing the desired shape while pushing down the syringe plunger, and thereby steadily pushing out the cheese into the desired shape). When melted, gooey cheese meets a cooler surface onto which the printer squeezes it, the system becomes solid (as, among other effects, the fat in the mix solidifies), and a shape can be built up layer by layer. Among our early successes was a cheese teddy bear and, when our paper was published describing the work, the response online from news sites and social media was amazing, demonstrating that (A) there is wide international interest in the technology and (B) a lot of people don't think that 3D printing teddy bears made out of cheese is necessarily the most useful way scientists could be spending their time!

Anyway, in this example, we were relying on the properties of fats and proteins in the cheese to build (or really rebuild) a structure that was already there, but into a more "interesting" shape. What if we deconstructed this completely and went back to more basic components? We now know enough about the properties of milk proteins, for example, to be able to dissolve them in water and add some salts and agents to control the level of acidity such that we can engineer a mix that at a certain temperature is a printable liquid (or paste) but, at a lower temperature, will set into a fine structure. We saw earlier in this book how some polysaccharides like xanthan gum also have this property, which could be used in 3D applications, as could the potential for exploiting other "tricks" whereby a liquid can rapidly or instantaneously made to turn solids, as when we add calcium to alginate, in a technique already very popular in molecular gastronomy.

These materials are all also notable for the fact that they are all about structure and not flavor, and so can be used like scaffolding to build a structure that has a certain texture or mouthfeel, to which we then add flavors, and maybe nutrients, to produce a huge range of different products. We could even picture the "pod" we put in our printer at home just being the structure-forming stuff, to which we could add whatever flavors we wanted, to experiment to our hearts content and finally discover what some things with the consistency of cheese but the flavor of chocolate and curry really tasted like.

This is getting close to a Star Trek Replicator for food, and indeed the comparison is often made, while 3D printers have indeed been flown into space for a wide range of uses.[8] The sky (or space) really is the limit here, in terms of what can

---

[8] Imagine a situation where a plastic part breaks, and, rather than need to have a spare, NASA could just email up a computer file describing the shape of the object, and an onboard printer could produce a replacement in minutes! For this reason, there have been apparently 3D printers on the International Space Station for a number of years already.

be pictured. Coming back to earth, we could picture vending machines where you selected the ideal properties of the most desirable snack that would address the parameters of your current attack of the munchies, in terms of texture, flavor, and even added vitamins and minerals, and then press a button and within a short time have your unique snack in your hand (and mouth). This would be like a version of a juice or smoothie bar where any combination of ingredients can be made into a liquid, but going far beyond into a huge range of imaginable solid foods.

We could also picture designing products with specific textural properties for different medical conditions, such as for individuals who had difficulty chewing or swallowing, or ideal non-messy foods for toddlers.

Overall, in terms of the future of food, 3D printing is certainly a hugely exciting area right now. Whether one could be found in every kitchen in 5 or 10 years, like a microwave or coffee machine, remains to be seen, and my instinct is that this might be a novelty niche ultimately, but the developments in this regard are certainly going to be fun to watch.

Final thoughts on the future of food processing

In this chapter, some further aspects of food processing beyond heating, cooling, drying, and the other most prevalent large-scale (and really also small-scale) processing principles, and in particular some of the newest technologies to be evaluated, have been introduced.

The ongoing evolution of food processing shows that this field is far from static, and new challenges arise all the time that may need new approaches for them to be successfully met.

For example, in the last few years, there has been a lot of discussion as to whether a bacterium that can be found in raw milk but that was previously of little concern, called *Mycobacterium avium,* subspecies *paratuberculosis* (sometimes abbreviated to MAP), could be somehow related to the transmission of Crohn's Disease. The reason for this was that the bacterium was linked to a somewhat similar condition in cattle, called Johne's Disease, and so the possibility existed that milk from cattle carrying that bug could perhaps contaminate human drinking milk, and that disease could result.

In such cases, the first question is whether the main line of defense we have in place to prevent transmission of disease through milk, pasteurization, is sufficient to kill this bacterium, because if it does that its presence in milk would not be of concern. Unfortunately, answering this question definitively was much more difficult than might have been imagined, principally because the bacterium in question was exceptionally difficult to get to grow in a laboratory, a requirement for determining if it was alive or dead after heating. Eventually, results of laboratory heating tests suggested it might be just a little tougher than the bacteria that traditional industrial pasteurization was designed to kill, and so in most plants the temperature of pasteurization, the time for which milk was held at the maximum heating temperature, or both, were extended slightly, to be absolutely sure that this new threat could be dealt with without question, if present.

The key lesson from this case is that, had we asked any food scientist up to this development which food safety treatment they reckoned we had absolute under-standing of and control over, it would have been pasteurization of milk. We knew it inside and out and upside and down and had absolute confidence in its efficiency in safeguarding the population from milk-induced bacterial illness. However, pride often comes before a fall, and the case of MAP shows that we can never be com-placent and that emerging threats we haven't heard of or thought could come along tomorrow and force us to rethink our processing principles. In particular, advances in detection technologies, where we can take a drop of milk, or any foodstuff, and identify through incredibly fast DNA sequencing technologies every bacterium pre-sent, is throwing up surprises all the time.

Put this beside the ongoing consumer demands for less processed, more natural food, and we can see where drivers for new approaches to food processing, or con-stant improvement of the techniques we already use, come from.

Consumer demands and views are in fact a huge driver in the development of food processing tools, as they should be. A technology can be developed and evaluated, found to be hugely powerful in solving specific problems with food, and yet never be deployed on the basis of consumer concerns alone.

Probably the most prominent example of such technologies is the use of irra-diation of food, in which food is exposed to radioactive rays either generated by specific machinery or simply exuded by naturally or artificially irradiated materials. I was once in a food irradiation chamber, a small room with ominously thick lead-lined walls and an enormous door on the wrong side of which at the wrong time you would not wish to be, in the center of which was a table over a trapdoor. Below that door sat a container of highly radioactive material, and when food was placed on the table and the room emptied of people, the trapdoor would retract and ghostly invisible rays of lethal radiation would wash over the food. This would kick off a series of reactions in the food that would highly effectively kill bacteria within, while causing minimal quality changes (once steps were taken to minimize the in-itiation of reactions like oxidation, like keeping the food frozen or wrapped in an oxygen-free package). In addition, irradiation has other applications, like inhibiting sprouting of potatoes, and is one of the only technologies that can effectively disin-fect dry materials like herbs and spices.[9] However, instinctive consumer distrust of anything to do with radiation means that this particular tool largely remains in the box unused, in most countries at least.

I haven't even mentioned all the new ways scientists are exploring to process our food and make it safer. Other techniques under study in laboratories around the world include using ultrasound waves to send shocks through liquids and hammer

---

[9] Indeed, in many European Union countries, irradiation is permitted only for dried herbs, spices, and seasonings.

bacterial cells to death, exposing them to incredibly bright light (the sunbed from hell) to fry them with light, or using something called cold plasma, which is a mixture of highly reactive and aggressive (if you are a bacterium, at least) molecules generated when we pass electrical discharge through a gas.

Overall, it is likely that in 20 years, when we look at the products on our shop shelves, the technologies that have been used to ensure our safety and give them a suitable shelf life will have changed or evolved further, while scientists will have gone on to other strategies and tactics in the never-ending war on the many agents that could cause problems with our food, from spoilage to illness.

14

Packaging of Food
SO MUCH MORE THAN JUST A BAG

Contain and protect: The functions of packaging

History and fiction provide many examples of barriers being constructed to keep undesirable enemies away, containing a few points at which crossing from one side to the other is carefully controlled, whether it be Hadrian's Wall, the Berlin Wall, or a 300-mile-long, 700-foot-tall ice wall.

This is also the basic principle of food packaging: keep the food and its quality in, and keep all things that can impair that quality, or threaten consumer safety, out. Food packaging has many functions:

1. Contain the product in a physical sense in a single mass and place.
2. Protect the product from physical damage.
3. Protect the product from chemical damage by controlling the access of gases, moisture, or light to the product inside, and also perhaps protect the product from sudden changes in temperature.
4. Present the product to consumers in defined portion sizes.
5. Provide convenience (for example, in how consumers can carry, open. or use and perhaps reuse the package).
6. Provide surfaces on which information (marketing and practical) can be provided to consumers.
7. Be compatible with whatever processes a consumer will subject it to, such as freezing or microwaving, if that is part of the function of the package and intended handling of the product.
8. "Be the least evil" in terms of impact on the environment and our planet.

The most common materials used for food today are cardboard, plastic, metal, and glass, while other materials occasionally used in food packaging include wood and ceramics.

To achieve the properties just mentioned, though, most food packages today are complex composites of several materials, frequently bonded together in a way that is invisible to the naked eye. Such arrangements are necessitated by the fact that no one material usually has exactly the right properties needed, so such composites (called laminates where layers are glued together, as of paper, foil, and plastic, for example) are used to give a combination of strengths that overcomes any individual component's disadvantages, like physical weakness, sensitivity to water or fat, or transparency.

Packaging also is complicated by the fact that for most products it consists of multiple layers at a macroscopic (visible) level. For example, in the case of a breakfast cereal like cornflakes, the primary package is the plastic inner bag in direct contact with the product but, to deliver the cornflakes to the consumer as flakes and not dust, which is surrounded by a strong outer cardboard container (the secondary package). This, being the package the consumer sees, bears the colorful label, cartoon characters, product name, manufacturer address, usage instructions, nutritional information, and such like. On leaving the factory, however, the product is likely to have been dispatched in much larger containers (tertiary packages) that batch together dozens of individual packages for sale to the shop or supermarket, and so forth. Thus, many food products typically are transported from point of manufacture to point of sale in a Russian doll–like arrangement of packages within packages to ultimately protect the product, protect the consumer, and provide convenience for all concerned

Today, the packaging of food is a hugely complex discipline, involving science, engineering, marketing, and psychology, to deliver the best container for the product, the consumer, and the environment.

A package of packaging materials

Probably the single most common material used in food packaging today, in terms of weight, is plastic, whether as the basis of a package in its own right, like a yogurt tub, or as part of a complex laminated material, as described in the previous section. Of course, there is no such thing as "plastic" (just as there is no single thing that is "food"), but the term covers a wide range of materials of very different properties. These all share the common chemical property of being complex polymers or chains of smaller and simpler molecules. The individual chemical building blocks of plastics are usually molecular structures that can readily be induced to link up into solid materials that can be rigid, film-like, squishy, or any of a range of textures and consistencies, depending on which plastic is present and what has been done to it to make it into a solid product.[1] Plastics thus have the advantages of the range of properties they can possess and the flexibility with which they can form different sized and shaped products.

---

[1] Most plastic materials used in food packaging are thermoplastics, which soften at high temperatures and return to their original form at around room temperature.

One of the main plastics used in food packaging is PET,[2] which is used in car-bonated beverage bottles because of its transparency, good ability to trap gases such as carbon dioxide, and such physical strength that it doesn't shatter easily. Polycarbonate is also commonly used to make drink containers, and large containers in general, while PVDC[3] is used in many prepared food packages, poly-propylene is used in bottles and trays, and polystyrene is familiar in packages such as egg cartons, fast food trays, and coffee cups (where its ability to entrap air is a great advantage in insulating hands from boiling contents).

Compared with plastic, glass is a material that protects products very well (except, unless colored, from that surprisingly damaging influence called light), and conveys traditional values, sturdiness, and stability. On the downside, it is heavier than an equivalent-sized plastic package, which can be inconvenient for consumers and food processors in terms of energy spent in transporting heavy packages as well as the food they contain, and is prone to breakage in an occasionally dangerous fashion.

Metal, as used for cans, shares the physical strength and strong barrier properties of glass, but is lighter and nontransparent, and stronger containers can be produced just by having thicker walls (compare a baked bean can with a beer can). Metal is also a good barrier to water, odors, and air, and a good conductor of heat, which is why it is the material or choice for products that were sterilized within the encompassing walls (and lid) of the package, including those canned foods of amusingly distant expiry date we occasionally stock up on for forthcoming emergencies or an impending zombie apocalypse. Tin cans are actually still partly composed of tin, found as a coating on a backbone of steel, or may be made of aluminum and steel coated in a lacquer layer (of plastic, to protect metal from food, and sometimes *vice versa*).[4]

As well as forming containers of sufficient strength to be shaped into a cylin-drical package, metals such as aluminum can be used in much thinner form as a foil, which can be built into a composite laminated package in which it can confer its light-impermeable, malleable, shiny, and gas-rejecting benefits while its relative weakness is compensated for by being coated with plastic layers.

Cardboard or paper-based packaging materials have the obvious great advan-tage of printability, being essentially a blank canvas on which labels, images, colors, and information can be presented, and they also keep light out while remaining light themselves. Card-based materials are also very flexible and can be formed into

---

[2] PET stands for polyethylene terephthalate. The prefix "poly" in the names of different types of plastic just means that they have loads and loads of molecules of what follows the prefix stuck together, in this case the snappily named ethylene tetraphthalate—if such chemical names freak you out, stop looking at the footnotes in this section!

[3] Polyvinylidene chloride—I did warn you!

[4] For some decades after their initial introduction as a food packaging material, around the start of the 19th century, cans were sealed using a tin–lead alloy. The joy of consumers at their now-sterile products saving them from bacterial food poisoning was somewhat undercut by the lead poisoning this caused, including possibly on some famous voyages of exploration. Luckily, they worked out that problem eventually.

nice neat geometrical shapes, like boxes and cartons, but suffer from a relative physical weakness and also a tendency to be attacked by common food components such as water and oils, leading to soggy mushiness, which necessitates their reinforcement with protective layers of plastics.

One of the key functions of all the types of packaging just listed is their ability to act as a barrier to something or other, whether gas, water, or light.

The prevention of light penetrating the package may often be due to the fact that deteriorative oxidation reactions are catalyzed by light and can thus simply be controlled by keeping light away from the sensitive molecules. In other cases, a nontransparent package may be desirable because the product shows some settling or other physical changes during storage, which can be readily reversed by the powerful instruction to "shake well before opening," but that might cause a consumer who sees solids or layers of material setting out of the product to conclude incorrectly that something undesirable has taken place.

Access of light and product visibility can also be controlled easily by having the label of the package completely cover its surface, if the product will still benefit most from being contained in a plastic material that isn't naturally opaque, or by using a nontransparent packaging material such as cardboard or aluminum foil. These are good light barriers and have good physical strength too, but in the case of card can lose strength rapidly when in contact for long periods with oil or water, while foil can be susceptible in thin layers to tears or punctures. This is where the concept of laminates come from, in that they are designed to compensate for weaknesses of individual packaging materials by having multiple layers of different materials, so that, for example, a protective barrier of plastic is placed between strong printable cardboard and the watery package contents that would just love to make that cardboard soggy and weak.

If a modern juice carton was dissected and analyzed, for example, the walls will be found to contain multiple layers bonded to each other in a highly defined order. These will typically include the following:

1. Cardboard, for physical strength and printability.
2. Aluminum foil, for insulation from rapid changes in temperature during handling of the juice, which could encourage microbial growth.
3. Plastic layers, to make the whole waterproof and also provide insurance against failure that is due to puncture of the foil or cardboard layers.

So, modern food packages are highly complex constructions at many levels.

Getting the atmosphere right

A major principle for the protection of product quality by packaging that is invisible to the consumer is the atmosphere present around the food. For some products, all the air is sucked out of the packages before they are sealed, and they are vacuum-packaged in tight wrapping like a mummy.

In such cases, the key to getting the atmosphere right is getting the atmosphere out, in recognition of the unhelpful role of oxygen in promoting reactions we would prefer not take place, including the growth of the significant proportion of types of bacteria that require oxygen for their survival.

Of course, air is good, and rather important to humans. We (and most other living organisms[5]) need oxygen to live, and get it from the air around us, into which we exhale carbon dioxide and water vapor. Plants also use oxygen, but also helpfully take in carbon dioxide and release oxygen through photosynthesis, and so the earth's atmosphere remains a finely balanced mix of what its populations need to survive and thrive.

Plants, animals, and bacteria have evolved to use this gas mix and do rather well out of it. However, when we convert such materials into food (especially plant and animal material) we want all the things that went on before that conversion, such as metabolism with all the associated changes that brings, to stop. As we have seen, we approach this goal directly through processing, for example, by heating food to stop enzymatic reactions; if we have a fresh food that we want to stabilize throughout sale, distribution, and home storage before it is processed (by cooking, for example) a powerful way of doing this is to manipulate the mixture of gases with which it makes contact. This is called modified atmosphere packaging (MAP), and MAP is the reason that many packaged food products are surrounded by what is usually assumed to be air but, in most cases, is probably not.

In MAP, the air is removed from the immediate environment of the food (usually the innermost packaging) and a replacement mix of safe inert gases is fed in. The key objective, as mentioned, is usually to reduce the oxygen content, which discourages many undesirable biological reactions, and of course some chemical ones as well, such as oxidation, which can lead to rancidity in high-fat food products, as discussed in Chapter 6. However, the objective might in some cases not be to remove oxygen completely, as its complete absence can lead to its own problems, such as loss of color in meat, or a different set of anaerobic reactions in fruit and vegetables that would spoil the product, just in different ways.

A further key consideration is keeping the gas mixture in the right state within the package, which depends on what gases the package allows to enter during storage and what can pass through the packaging material to the air outside the package. The packaging material, and its permeability to gases, is thus a hugely important part of getting MAP to work right.

---

[5] We are aerobic organisms, and not just when clad in Lycra and trying to get fit. Being aerobes means that we depend on oxygen for our existence. We share this with the vast majority of living things on earth, from trees to terrapins to terriers, and many microorganisms, but some bacterial species actually thrive best in the absence of oxygen, which they regard as toxic. These species are called anaerobes; some other bacteria hedge their bets and can survive with or without air, and are called facultative anaerobes (swimming competitions, among many other things, would be very different if humans worked out this trick).

To make things even more complicated, for products that essentially continue to respire (breathe) inside the package, like fruits and vegetables, it is important that the package have some very limited degree of permeability to gases to get the right balance of atmosphere within the package, as otherwise the oxygen in the package would be consumed by the respiring contents and quality problems would ensue. The package in this case must mostly keep in the MAP gases while allowing some oxygen to leak in to replace that which has been used up.

For other foods (like cheese and meat), the material will usually be chosen for its complete prevention of gas movement in or out of the package. The range of plastic materials available for food packaging luckily provides a menu of options with different gas barrier properties that can be chosen from, depending on the product needs.

Besides adjusting the oxygen level to strike the right balance between being good to the product and hostile to unwanted biological or chemical phenomena, MAP strategies often involve increasing the level of carbon dioxide, taking advantage of its ability to inhibit bacterial growth. The remainder (indeed often the majority) of the atmosphere, after the right levels of oxygen and carbon dioxide have been calculated and dosed in, is nitrogen, which does little directly, being spectacularly unreactive and boring, but dilutes its more powerful package-mates to the levels where they can best exert their desired powers.

So, the particular gas mix chosen will depend on the particular product concerned and what is to be preserved and prevented during storage. For fresh meat, for example, we want to maintain a strong red color, which means keeping a protein called myoglobin in the meat complexed with oxygen (without which it would turn, and the meat itself in turn would turn, an unappetizing grayish-brown color), and the MAP mix must be quite rich in oxygen; this of course creates conducive conditions for many bacteria, so must be accompanied by a high level of carbon dioxide, which does nothing to the color but prevent the bacteria from getting carried away in the high levels of oxygen present.

The case of fruits and vegetables has been already mentioned and, if these are cut, oxygen becomes a major concern, as it is acted on by an enzyme called polyphenol oxidase (PPO) in the fruit. The ensuing reaction causes the brown color and softening of texture associated with fruit surfaces that are exposed when the fruit is cut or damaged. In these products, the key will be to get the level of oxygen down and replace it with carbon dioxide.

In the presence of oxygen, PPO converts polyphenols to compounds called quinones and eventually (in the presence of oxygen) to melanin, which is a common brown pigment found in many different systems (including our skin). Normally, the PPO and the phenols with which it can cause such undesirable reactions are kept separate by the architecture of the fruit, with cell walls and structures within the cells keeping them apart. Attack nature's finest work with a knife, however, and things that were not meant to mix do so, and discoloration ensues thanks to PPO's oxidative rampage. The mixing doesn't have to end in unpleasantness though, as the conditions for the reaction can still be manipulated to slow down the progress,

for example, by keeping oxygen out of the mix (through the gas mix manipulation previously mentioned) or else adding something that discourages oxidation.
As mentioned elsewhere in Chapter 6, lots of food ingredients and additives are
natural antioxidants, but in the case of fruit a particularly compatible one might be
the ascorbic acid (vitamin C) in lemon juice.

In the case of fish, we need to consider whether the goal is to avoid oxygen that
would cause off-flavors through oxidation of high-fat fish or to discourage aerobic
or anaerobic spoilage bacteria, and so we need to balance how important oxygen-
dependent reactions are compared to color preservation or deterioration. So, for
different types of fish, different mixes with different levels of carbon dioxide or
oxygen will be selected.

Then, for products such as potato crisps (or chips in the United States) the package
is usually filled with almost purely nitrogen, which serves two functions: preventing
undesirable changes such as oxidation[6] and giving a nice puffed-up "pillow" in
the package, which cushions the fragile contents within from bumps and impacts
during transport and hopefully delivers nicely shaped intact snacks rather than a
bag of crumbs.

All in all, when you next pick up a package and think it is just full of air, be
aware that it is very unlikely to actually be air, and more likely to be a carefully
selected mix of gases that have done a remarkable job getting the product to you in
hopefully an optimal state and that you then release from service into the wild when
you cut, snip, or burst their prison.

Packaging and perception

It is known that subconscious psychological factors have a massive impact on how
we regard food packaging and the conclusions we will draw about a food product
based on how it is presented to us in a package. Numerous studies have shown that
consumers will respond to the same product in very different ways just based on the
package in which it came.

This obviously places a huge importance on the correct selection and design of
a package for any food product, because the package is the first point of contact
of consumer and product, and chance for the product to make a favorable first impression. For a new product, for example, the goal must be for a consumer to see
and be attracted to the product on the shelf and make a decision to buy and try it.
Packaging is the first step in being successful in this regard, and the next critical step
(of tasting and liking the product within) won't even happen if the initial decision
to purchase wasn't successfully induced.

---

[6] The low-moisture content will pretty much take care of bacterial growth, but the presence of oil
in the formulation or coming from the frying process used during manufacture can present a target for
oxidation.

Careful choice of color schemes can have a great impact on perception as, for example, lighter colored packaging is associated with healthier foods. Conversely, there is evidence that light-colored packaging can make consumers expect a less tasty or rich food, even though they might perceive it as healthier. Green colors are perhaps unsurprisingly associated with natural products, while blue gives a much cooler and calmer impression than red. Red is in fact associated with sweetness, and in recent years consumers complained that Coca-Cola had become less sweet when the manufacturers simply tried a white rather than red can. Such an association has now been clearly scientifically established, as experiments at the University of Oxford showed that popcorn served in a red bowl was deemed sweeter than the identical product in different-colored bowls.[7]

Package shape is also very important. There is a lot of evidence that foods presented in angular packages with lots of corners and edges will be perceived to have a more intense and strong taste when compared with something presented in a rounder package. This can be demonstrated with something as simple as consumer scores for flavor intensity of a yogurt, which will be measurably different in a rounded pot than a square one.

Retailers can also play with psychology on a larger scale in terms of how a food product is presented for sale, including factors such as lighting, the presence or absence and type of music, closeness to eye level of presentation, position within a shop or market, and closeness to the exit and tills. A modern food market has been designed to within an inch of its life to exploit the unspoken and unwilling psychological biases and tendencies of the shoppers walking through the door.

So, we are constantly being subtly (and sometimes not too subtly) being influenced by carefully designed food packages that manipulate our minds and choices based on understanding of human psychology. Before you get too indignant at such interference, however, just remember that nature has been at this trick a long time before the food industry discovered it, as, for example, when the ripest fruits have the most vivid and attractive color, the better to get the parent plant's seeds dispersed.

Active and intelligent packaging

In the examples described in the previous section, the key consideration was designing a package that is perfectly suited to maintaining the quality of the food and that did the same job in protecting and maintaining that from the day the product was put into the package to the day when the package was opened for consumption of the goodies within.

---

[7] The relationship between food packaging and psychology, as demonstrated in such experiments, is very nicely explored in Professor Charles Spence's book *Gastrophysics*.

What if packaging went a step further and somehow responded to or reacted with the food during its storage and shelf life, to dynamically control the food quality or perhaps provide information to the consumer about the quality or safety of the food therein?

Some years ago, an enterprising young Irish student won a national young scientists' competition for her brilliant idea of making a sachet containing bacteria, sugar, and an indicator that changed color when acid was produced, which was then put on the outside of a milk carton. If a typical carton was left out of the fridge (for example, if a delivery was left outside a shop on a hot day for a few hours before being picked up and put in the display fridge), its appearance will not betray this nor the increased dangerous risk of bacterial growth within. With the sachet present, however, leaving the milk out would warm it up sufficiently that the bacteria would grow, turn sugar to acid, and a color change to red for danger would result. This is such a simple and fantastic idea that I am amazed it hasn't appeared on every carton of pasteurized milk since, and it shows the potential for packaging to go beyond a simple (or even not-so-simple) containment device to something much more interactive and instructive.

Imagine if, in ways such as this, your food package spoke to you, verbally or visually, and told you that the product was not safe or needed to be dumped? It is not hard to visualize this being a reality in the not-too-distant future, and it is also predicted that such automated sensory perception powers might be incorporated into fridges and other kitchen devices too.

Such packaging approaches are referred to as intelligent or active packaging, for reasons that make sense, but should not be taken as suggesting that traditional packaging approaches are dumb or inactive!

Another example might be a package that responded to changes in the food over time and worked actively to mitigate these if those would otherwise be undesirable. This might involve, as sometimes already seen today, a package containing a small sachet (typically marked clearly as not to be eaten!) that contains a substance that absorbs undesirable gases or odors emitted by the food or controls the oxygen level within the package. The negative impact of oxygen could also be controlled by incorporation of antioxidant substances into a package, for example, embedded in microscopic beads into the surface of the package closest to the food.

In other approaches to active packaging for meat, the key is the release of carbon dioxide during storage, which (as previously mentioned) helps to inhibit bacterial growth; such emitters can be activated by the natural release of some moisture from meat or fish during storage. Too much moisture release and mobility in food is a problem though, because of the key role of water in so many undesirable reactions in food, and stray moisture could be controlled by placing a simple absorbent pad in the package, as often seen today underneath a piece of refrigerated meat or fish.

A package could also be impregnated with a substance that inhibits bacterial growth in such a way that it is gradually released during storage and serves a function equivalent to a security guard patrolling a premises, keeping unwanted intruders at bay. Several such systems have been developed based on the ability of silver to kill bacteria, and silver-impregnated materials can be used to keep the surfaces of food in contact with that material free of bacterial growth, while other systems use materials that can diffuse through the food (perhaps as a gas) to prevent undesirable growth.

Packaging can also help with the processing of the product by the consumer, for example, by including a metallic component that heats in a microwave in such a way as to achieve browning and crisping of the base of a pizza slice that otherwise could not be rapidly heated for convenience without looking pale and limp.

In terms of intelligent packaging, the approach described earlier that involves the use of bacterial growth in a special "sensor package" to tell if a product has been exposed to a dangerous incidence of what is called temperature abuse is an example of what is called a "time-temperature indicator," or TTI. Other examples of this might be changes in color based on a chemical reaction that requires either a long time or exposure to an unacceptable temperature (or both) to take place. Indicators might also involve reactions that are triggered when a package is open, and then proceed to change in some way like a progressive darkening of color, so that consumers can be reminded of how long it is since a package was first opened.

Sensors have also been developed that detect with exquisite sensitivity the presence and level of particular molecules in food, where this reflects the state of freshness or spoilage of the product. Some sensors can even detect and report the presence of specific individual bacteria of concern, and in all of these cases the detection of the unwanted substance or microorganism results in an easily determined change such as a color change or darkening of an indicator strip.

To add an extra layer of technological sophistication to such processes, such sensors could become tags that can communicate with external computers, monitoring the quality and transmitting updates to the retailer, processor, or consumer as to what was going on in the package. This can also be linked to goals to improve traceability of food products, such that the origin and location of products throughout the food chain from production to consumption can be tracked and safety and integrity ensured on this basis. It isn't too difficult to extend this line of development to an app on our phones,[8] through which we can check in on the freshness of what is in our fridge, to determine what needs to be disposed of, and what needs to be added to our shopping list for replacement. This type of approach is already being seen today.

This kind of approach can be seen to have huge potential to reduce food waste, a major and increasing concern in many countries, as decisions as to whether to consume or discard food would be made on the basis of actual evidence of food quality, rather than "best-before dates." It is known that many consumers interpret these dates as reflecting a point beyond which a product is no longer safe, rather than relating instead, as they do, to food quality, and the fact that many foods can be perfectly safe and acceptable to eat beyond those dates. Intelligent and active

---

[8] Perhaps called "Spoiler Alert"

packaging can then provide real-time information that adds dimensions of detail to these dates by telling us exactly what is going on in this particular batch of this particular product, which has been subjected to this set of conditions of storage and handling, and whether these are more or less ideal than those for which the original best-before date was estimated.

The difference between active and intelligent packaging is that the former proactively works to control the safety or quality of the food, while the latter might not directly do so, but provides information about this to the consumer, retailer, or processor. Packages that combine both intelligent and active characteristics and functions are described as "smart" and are likely to be increasingly dominant elements of the future of food packaging. Packaging has certainly come a long way from being just a bag or box.

The ultimate in reducing waste: Edible packaging

With increasing awareness of the impact of discarded packaging, whether cardboard, plastic, or paper, on the environment, the biodegradability of packaging has become a major concern for the food industry, consumers, and concerned groups alike. The selection and design of food packaging today must thus have as a primary consideration the minimization of material use, the ability of materials to be reused or recycled (and the provision of clear information to consumers on this), and the overall least possible negative impact on the environment, in terms of factors such as wastage, dumping, and the cost and impact of recycling materials.

One interesting way to approach such minimization strategies is to make the package itself, or part of it, edible, and itself a food commodity. In a way, this is simply following the example nature gave us with the peels of many fruits, for example.

There has been a lot of research in recent years into how structure-giving materials we have met earlier in this book like proteins and polysaccharides, or mixtures of these, instead of giving three-dimensional gels or structures could be induced to form almost two-dimensional flat sheets (the third dimension, thickness, being very small) that could then be wrapped around a package. Key properties for any such material would be those applicable to any other package, like strength (we need the film to fold and bend without tearing) and control of movement of potentially deleterious agents like moisture and gases. Biological stability would also be key, in that the package itself wouldn't be a target for bacterial or mold growth, for example. The package, once no longer needed, should then be consumed or be allowed to compost rapidly.

One example of a material much studied for its use in edible packaging is called chitosan, which is a polysaccharide widely found in nature, for example, in the shells or crabs and other crustaceans and as a structuring material in fungi. Chitosan can be readily formed into films and is a good barrier against gases and light, with the added bonus of having antimicrobial properties, and the idea of a material tough enough to protect the soft insides of a crab protecting your food (albeit in a much thinner form) seems like a good principle of natural preservation to me. Other ingredients can also be added to the film material before it forms to enhance the protective or nutritional function of such packages.

Proposed applications for such packages include individually wrapped fruits or pieces of meat or fish. The films could also be used inside products, where regions of high moisture and low moisture need to be kept apart, to prevent moisture migration and softening, for example, between a pizza base and its toppings, or the pastry crust of a pie and the creamy innards within. There have even been developments of edible tableware, including knives, forks, and plates.

A recent development in this regard is called WikiPearls, which consists of an edible shell or skin that can be wrapped around food like yogurt, which it protects like a traditional package but can be eaten in its own right, unlike (under normal circumstances!) a traditional package. Kentucky Fried Chicken even proposed to introduce edible coffee cups (the "Scoffee" cup!), made from a biscuit covered in sugar paper lined with a white chocolate layer, such that the heat gradually softens the biscuit as the chocolate melts, and as the last sip of coffee is downed the cup can follow swiftly thereafter;[9] the cups could even be infused with different evocative aroma combinations. Other examples include edible gelatin-based beverage cups, cupcake wrappers made from potato starch, and rice-paper-based films which have been used in Japan for many years.

Challenges remain though in making such approaches economical, given the cost of ingredients and manufacture. Nonetheless, this remains a very active area of packaging research, and it seems highly likely we will increasingly encounter food in packages where waste is effectively eliminated by the fact that, when finished the contents, we can tuck into the wrapper for dessert.

---

[9] Perhaps inspired through pure imagination by the teacups in Willie Wonka's Everything Edible Room.

15

Innovation and the Development of Recipes and Formulations

What is innovation in food?

Many studies have reported astonishing statistics about the rate of introduction of new food products globally, with new products appearing probably at least every hour somewhere around the world, if not more frequently.

If you could go into a food store anywhere in the world and somehow take a snapshot of the range of products on the shelf, then revisit it five years later and do a comparison of what you find, there would be a huge surprise in terms of the turnover. Many products will have disappeared, and many new ones will have appeared. For those that remain across this time span, there is a very high likelihood that they have changed in less visible ways, in terms of modifications to their formulation, package, or the process by which they are made. Even fresh foods like fruit, vegetables, meat, and fish are likely to have benefited from scientific advances in their production, quality, or transportation in an optimal state of quality and safety.

Why is there such a high rate of change? There are two main drivers, one external to those who produce the food and that relates to the highly fluid and sometimes unpredictable expectations and demand of consumers, and one more specific to the food producer that relates to new opportunities in technology, formulation, or scientific understanding.

For any new product to be successful on the market requires two successful changes in behavior of consumers. The first is that, instead or as well as what they normally purchase, they need to buy to try a new product, and drop it into their basket or cart as a result of a planned or spontaneous decision to do so. To achieve this is primarily the responsibility of experts in marketing, who can divine what consumers want, develop a strategy accordingly, and then deploy the appropriate tools to bring the product to the attention of those who are most likely to buy the product, such as through promotions, advertisements, and probably, in today's world, social media campaigns.

The second behavioral change that a product must achieve is that, having bought the product once, consumers must want to continue to do so, because they like what they have found. This is firmly the responsibility of food scientists, who have delivered a product with the expected and optimal flavor, texture, stability, cooking properties, or whatever the taster was led to expect by the wiles of the marketers.

Studies of food product innovation also consistently come to another conclusion, which is that, for the vast majority of the new products that have appeared, there is actually nothing new about them.

Many new food products that appear on the market are clones that follow in the food-steps of other successful products, and hope to achieve some of the reflected glory of their more successful predecessor. So, every time a new product appears on the market that captures public attention, it is only a matter of time before clones appear, which frequently have as their primary differentiating point a cheaper price. In such cases, however, Darwinian evolution kicks in with a vengeance, and only the fittest will survive, so that within a year most of the clones will have disappeared from the shelves, unless one has managed to usurp the original product and has clambered to the top of that category tree.

There are many examples of this type of phenomenon outside the world of food, for example, in consumer electronics, as when each time Apple has brought out a new device over the last decade it is followed by imitations, which frequently fail to offer any substantial benefit over the original that they seek to emulate. So, the iPod was followed by a slew of large-capacity portable electronic music players and the iPhone by a horde of touchscreen phones, while the iPad led to the emergence of a whole new category of tablets. Some of these that have followed have been good and have succeeded in eating into Apple's share, but most have disappeared from sight after failing to offer consumers a compelling reason to prefer them to the original product.

Indeed, when such a new product appears, a process called reverse engineering begins, whereby competitors literally and metaphorically take apart a successful new product to work out how it was made, why it is a success, and how to emulate it. Such practices can be found in every type of manufacturing industry, and food is no different. For these reasons, finding a true original can be a very rare experience, and hence a very much sought-after one, whereas the experience of encountering *clones* of comparable products is far more common.

If we look at products that do not start with a blank sheet where we invent something new, but the products are not copies of something preexisting under another name, there are several categories of product with which food producers can attempt to tempt consumers. I am going to consider these in turn, and then discuss how these principles apply, whether it is a food company making something new to be sold directly to consumers or a chef coming up with a new meal in a restaurant.

The first, and perhaps simplest, category of new food products that might appear on the market, and one that is very common, is called *line extensions*, where a company takes an existing product of its own and makes a minor modification, typically in terms of flavor, to create a sister product to the original. An example of this might be a product like crisps or potato chips, where new flavors regularly appear on the shop shelf, some of which might succeed and stand the test of time, while others disappear if consumers don't like the new flavor. Such developments, from a food production perspective, are technically and scientifically pretty straightforward, as most of the parameters that describe the product (the main elements of the formulation and most of the production process) remain unchanged, and only one attribute must be changed or optimized. Other examples might be new flavors of soft drinks, minerals waters, or yogurt.

The complexity of even such a minor tweak should not be underestimated though, as every new ingredient brings decisions and evaluations, such as the source (Where will it be found? What is the cost? How reliable is the supplier? How far away are they? Can they produce enough for what we need? Is their ingredient of good quality?), point and method of addition within an existing production line, and level of addition (too little might mean no flavor is detected, too much might be too strong, and every added molecule effectively adds cost.

Moving beyond these simple line extensions would be *reformulations*, which take an existing product and modify it in far more fundamental ways, for example, taking those chips or snacks and reducing their fat content, making them gluten-free, or making a version that is based on rice instead of cereals. While the flavor modification could be said to be (literally) a surface change, this is much deeper under-the-hood reengineering, as removing one component, with or without adding a new one instead, can have fairly fundamental effects on the characteristics of that product, and getting a new version of the product that matches consumer expectations can be far from a trivial task.

Many of the drivers of reformulation are nutritionally based, such as the need to match consumer desires for products that are healthier, or at least "less evil," by reducing their fat or salt content. Such cases are often complicated, though, when you consider why salt or fat was there in the first place and what they contributed to the product (flavor, preservation, and control of protein functionality in the case of salt, and texture in the case of fat). These considerations mean that removing them can have knock-on consequences, which mean that producing a superior product in their absence is very difficult. For example, despite decades of effort, hundreds of scientific papers and probably dozens of PhD theses around the world, there is still arguably no low-fat cheese on the market that a consumer would taste unprompted and mistake for a full-fat equivalent.

Innovation can also come from presenting *old products in new forms*, such as a form that is instantized, frozen, miniaturized, dried, microwaveable, or otherwise reinvented in some fundamental way. The objective here should normally be to offer consumers some labor-saving convenience advantage, which overcomes a deficiency in the old product of which they might not even have been aware.

Innovation can also come about through combining existing food types of components in new and hopefully interesting ways, as when yogurt and fruit were first paired together in divided packages, or when ready meals or dinner kits combined all the elements that are needed for a single meal, but would otherwise have to be purchased and prepared separately, into a single package and single means of preparation for consumption.

Another category of food product innovation, which has the advantage of frequently requiring no scientific or technological input before bringing it to market, is called *repositioning*, where a product is effectively sold in a new way, to draw attention to an attribute that wasn't always foregrounded in the way in which it was portrayed to consumers. For example, a company could decide that the fact that their product happens, and perhaps always happened to be, gluten-free, low-fat, or low-salt, is something that should be shouted about in a way that was not previously the case. The product is effectively relaunched, often without any modification except to the package and marketing approach, in such a way as to fore-ground its newly found virtues. This can also happen when the public perspective or scientific understanding of an ingredient changes, as when it became accepted that soya-derived ingredients had benefits for heart health, and products containing such ingredients could be now positioned as being heart protective.

Related to this, one of the most simple ways of presenting consumers with some semblance of "newness" is to redesign the *package* of a product, changing logos, colors, shapes, or images, so that the eye is drawn on a shelf to the appearance of something new, but where the contents of the package remain identical. Of course, a change in packaging can be more fundamental than this, whereby the material or construction of the package has changed in such a way as to better protect and preserve the product within; advances in the science of packaging have led to substantial improvements in the shelf lives of many products over the years, as discussed in the previous chapter.

With any of the types of innovation just described, a key question is also whether the consumer sees, or is supposed to see, the change. It might seem surprising to think that a company might change their product or reengineer it in a way that is invisible to the consumer, but minor tweaks to formulation, ingredients, or package might not be intended to gather new consumers, but have other goals, such as offering a longer shelf life (allowing the product to be sold farther away) or commonly to reduce costs or increase production efficiency.

Fear of failure and why food companies need to face their fears

All food product development involves risk. Many studies have, after exhaustive analysis, reached the depressing conclusion that the vast majority (probably more than 90%) of food products launched onto the market disappear without trace and are not to be found within perhaps a year of launch. The key reason for such failures is typically a lack of a compelling argument to consumers as to why they should purchase the product or a failure to deliver what was promised, in other words failing on the two key goals outlined earlier. Companies can also underestimate or overestimate the potential demand for a product, or launch products that confuse, disgust, or bemuse the consumer, occasionally all at the same time.

No company, no matter how large or how famous, is without their failures in this regard, as products like Crystal Pepsi and Guinness Light will attest. However, the risk of failure, wasting money and effort, and possible damaging consumer perception of a company or product brand, are arguably very different for a highly successful company that has many other sources of revenue and activity than for a small company or entrepreneur launching their first and only product onto the market.[1]

Key considerations might also revolve around the investment required to produce the proposed new product. Will a new piece of equipment be required, or even a whole new line? For a company making product A who wants to introduce product B, will they use the same process line (thereby either running it for longer or displacing some production of A), or build a new second plant or line? Will they need to buy a new packaging machine or system? The larger the investment decision and potential impact on production of other products, as well as personnel requirements, the stronger the case will need to be. What would happen if, in a year's time, the product was a failure and a new set of equipment lies idle, or else it is a far greater success than expected and the scale of production required to match demand exceeds the capacity put in place to make the product?

It there was a secret to foolproof product development, surely it would have been found by now, and products would no longer fail, and this is clearly not the case. It is widely accepted, though, that, for product innovation to have the greatest chance of success, a key factor is a close harmony between those who talk and listen to consumers (the marketing side of a business) and those who develop the products to match what those ears hear (the science and technology side).

Another key factor is disciplined process management from idea to launch, such that at each stage of the process success or failure is ruthlessly evaluated, and concepts or products that are not heading toward a likely successful future are abandoned or else sent back for further work. Good product development should look like a funnel, into which a large number of ideas are fed at one end but that progressively narrow as cruel Darwinian evolution winnows out the least successful options and keeps in play only the most promising strands of activity. The great biologist Linus Pauling, when asked how he had so many good ideas, replied that "I have lots of ideas, and I throw away the bad ones."[2] This should be a mantra for

---

[1] Among legends of food product failures, it is often reported that the slogan claiming that a beverage "brings you back to life" was translated in some Asian markets as the rather more alarming "brings your ancestors back from the dead," while Kentucky Fried Chicken's slogan "Finger-licking good" was similarly mistranslated as an encouragement to bite off your own fingers. Clearly, getting the message correct for the target audience is a key element of bringing a new product to its target audience with the intended impact.

[2] Pauling is probably my scientific hero. Among Pauling's particularly good ideas concerned the nature of how atoms bond with each other in molecules, leading from simple molecules to his later unraveling of some fundamental aspects of protein structure, as well as the idea that nuclear bombs were not a good thing. For these passionate causes, he was awarded two Nobel Prizes, for chemistry in

food companies, but of course the secret is knowing which are the good ones and which are the bad ones, and then being hard-hearted enough to kill a bad idea that might be a cherished pet product of someone, which he or she has worked on for weeks or months, but which initial sensory analysis, for example, suggests clearly is not going to be accepted by consumers.

It should be noted that of course there are limits as to what food companies can bring to market. In the EU, the United States, and elsewhere, there are strict regulatory controls on what can be launched as a new product. In particular, there are categories of new product for which specific approval must be sought before it can be introduced onto the market, leading often to years of research to demonstrate that a product is safe. Examples of innovations that might fall into this category include use of a new ingredient or raw material that has no history of safe use in the area under regulation (say, for example, if a new fungus was discovered in an arboreal forest and proposed to make a rare and unique truffle), or of a technology that has not previously been shown to have no adverse effects on the food, or its package where relevant. In the latter category, in the 1990s, a number of companies had to go through a highly complex process to demonstrate that high-pressure processing, discussed in Chapter 13, was safe to use as a process for food treatment in the EU. After that, each time a new product category was subjected to that treatment, lesser but essential evaluations would be undertaken to confirm that those products were also not negatively affected by the process. Other examples might be the use of a bacteria or other microorganism in food production that had not been previously used or a proposed introduction of a genetically modified organism.

A key concept in such evaluations is the determination of whether the new or newly treated food product is "substantially equivalent" to a product on the market that is being safely consumed on an ongoing basis, across a wide range of criteria, from microbiological and chemical to toxicological and nutritional.

This means that a company who wishes to bring forward something radically new (the term "novel" is often used for such products, to differentiate them from the simply "new"[3]) must accept that to do so will entail embarking on an expensive and exhaustive process, likely to take years, before it starts to make money. The market research and scientific basis for making such large investments must be very strong in such cases, as the risk is proportionately somewhat higher than making a simple line extension to an existing product. However, from a consumer viewpoint, having such safeguards in place is a very reassuring step in ensuring that safety above all else is a key criterion that has been rigorously evaluated before they pick up an

---

1954 and for peace in 1962, making him the only individual to win two Prizes in such diverse spheres of interest. It must be admitted that he, later in his career, failed to work out the structure of DNA, and proposed a triple-helix structure that was rapidly displaced by the correct double-helix model of Watson and Crick, so even his idea filter wasn't infallible.

[3] What are called Novel Foods Regulations, in the EU, are not to be confused with Food Novels, which would be much easier to read.

apparently exciting new product off a shelf and wonder about tossing it into their basket or cart.

Innovation in the kitchen

Any food product can be described as a set of ingredients and raw materials, subjected to a particular series of processing steps, and then put in a package and stored under conditions in which it should remain safe for consumption and present an acceptable appearance, taste, and quality to the consumer.

How does this relate to innovation away from industrial production, though, and in the context of the kitchen (domestic or restaurant)?

By analogy to the summary of what any food product on a shelf comprises, as previously summarized, so any meal on a plate is the sum of a set of ingredients and what has been done to them. Open any recipe book, and each recipe consists of two parts, the list of ingredients and their quantities and the operations to which they will be subjected prior to eating.

So, where can and does innovation come from here? Let's look at the types of innovation previously discussed.

What would a line extension in a kitchen be? Arguably every restaurant that offers a range of toppings on a burger or pizza is offering a set of extensions of a single line, and every new combination or topping added to the list constitutes a line extension. On many menus, the products will be familiar from other circumstances or restaurants, and old staples will appear again and again (familiarity leads to demand), and such standard offerings could be described, without disparagement, as clones.

On the other hand, if a restaurant changes beef in their burger for lamb, or a vegetarian substitute, or makes their pizza base gluten-free, they have reformulated.

Is there a packaging equivalent for a restaurant? I would argue yes, as in this case the packaging consists of the room and the manner of presentation of the food in question. Change the size, shape, color, or material of the surface on which the food is served (I am thinking here of wooden boards, slate plates, square plates, white or colored plates, small metal buckets, or even small cardboard boxes), turn up or down the lights, even change the type and volume of music, and the package of the food has changed and the impact of such "nudges" on our perception of exactly the same food product can unquestionably change.

So, where does true innovation in a kitchen setting come from?

In a food product, as we have seen, the most likely sources of real innovation might be discovery or use of a new ingredient or processing technology, the former presenting a new experience for consumers, while the latter might offer a new format or structure. Other powerful innovations can involve combining familiar ingredients and processes in new ways, as already discussed.

Even a quick consideration of why restaurants become successful, or what the top restaurants in the world offer to their diners, shows the chance for adulation through offering new or different experiences to that which can be encountered elsewhere, often based around exactly these pillars of ingredients or technologies. For example, a major trend in the revolution in Nordic cuisine in recent years has been the sourcing of unusual ingredients, such as those found by foraging or even the consumption of dishes based around insects, while many top restaurants all around the world are recognized for the ability to offer customers experiences and items on the menu that they can't get anywhere else.

In addition, a key element of innovation in the most exclusive restaurants is the use of unusual technologies, such as *sous vide* cooking, foams, freeze-driers, smoking, spherification, and other methods found in particular in the kitchen laboratories of proponents of molecular gastronomy, as discussed in Chapter 17. Again, taking new pieces of equipment or pieces of equipment previously used for one purpose and repurposing them for another application is an area of innovation practiced by processors of food at both large and more intimate scales.

Finally, kitchen innovation can come from the combinatorial interaction of novelty in ingredients with novelty in processing application, and also of the senses to be manipulated to generate excitement and satisfaction. As we have seen, the brain can be a powerful target, as psychology is played with to deliver unexpected flavors in atypical packages, such as food products that look like one thing but taste like something quite different and sufficiently unexpected as to elicit a gasp on first taste.

Interestingly, in considering the similarities between innovation as practiced by food scientists and chefs, which show them once again to be practitioners of equivalent skills and underlying principles, there is one area where the two worlds are actually directly linked in modern food product development. Today, in many companies and organizations, the initial development may be undertaken by chefs who either produce a new product from first principles, as it were, or else demonstrate how a company's ingredient can be used in a particular interesting food application. So, the worlds of the chef and the food scientist increasingly come together at the initial step of defining a food concept to describe what, under perfect circumstances, the experience of the consumer with that product or ingredient should be.

Assuming the end target is a packaged product designed for sale, though, this step is then followed by what is often a very difficult and unpredictable process called *scale up*, whereby food scientists must figure out how to take what has taken a chef perhaps hours of careful manipulation and attention to create and work out how to make 1000 of those per day, or 10,000, in a labor-efficient manner that makes economic sense and yields above all a consistent product, such that each of the replicates coming off a production line shares identical characteristics and quality.

The process of scale up usually covers several steps from development kitchen to production plant, with a key intermediate being a pilot plant, where equipment to be used in eventual production is available on a much smaller scale, so that more combinations and test batches can be evaluated without the cost and disruption of allowing such tests to be done on a much larger production scale. The formulation and processing conditions can then be optimized in the pilot plant, giving much greater confidence that, when transferred to tentative larger-scale production, success will result. However, every phenomenon involved in making or handling a food, from mixing to heating and flowing, as well as rates of key chemical reactions, can behave differently in different pieces of equipment at different scales, and what works for 1 kilogram (kg) can be very hard to make work for 10 kg, and impossible to replicate at 1000 kg. In addition, steps that are very handmade and delicate can be extremely difficult to scale up and automate, such that the hands of the chef can be mimicked by a set of processing equipment trying to do what those hands could do, but many times over.

Above all else, it must be noted again that a key requirement is that companies be ruthless throughout the process and be prepared to admit when the results of the scale up, or the initial consumer reaction to test batches of the product, are not pointing toward the likelihood of a successful outcome. Each stage of the process should be followed by a "gate," at which a hard look is taken at initial results and a decision made as to whether the product should continue to the next stage of development, be abandoned, or perhaps be returned to an earlier stage to modify the formulation or process so as to fix the problem which has been identified by the stern gatekeepers. I suspect each failed product could afterward be analyzed to see at what point signals were clearly visible that flagged eventual problems but were ignored or overlooked.

Developing recipes and formulations

If we open any cookbook, every recipe can be broken down to two pillars of instruction: the recipe and the preparation instructions. Similarly, every process can be summarized in terms of the inputs and the steps then applied to these to give the desired output.

Every food product or dish starts with the recipe or formulation (the term we might use for produced rather than kitchen-produced products), and this is really the definition in many ways of what the food is. The focus of this section is on how formulations and recipes are developed, whether in the high-spec environment of something like a modern infant formula factory or in a kitchen.

The key consideration in preparing any formulation or recipe is thus, to start by stating the obvious, to decide what goes into it. What kinds of factors might be considered at this stage?

1. What function will it serve in the food (are we adding it to contribute something to the nutrition, color, taste, structure, or to preserve the food, for example)?
2. Where will I get it and is that supplier reliable? If I have multiple options for where to get it, what criteria will I use to make the choice among these? What does it cost, and how important is cost as a criterion?

3. How much of it needs to be present (closely linked to question 1—add too much and it might be negative, add too little and it is useless—and question 2—the more we add, the more it costs)?
4. How will it interact with other ingredients in my product or dish (e.g., an acid might change the solubility of a protein, as might a mineral)?
5. Will its presence cause any consumers of the product concerns and hence reduce the pool of potential happy consumers (for ethical, dietary, disgust reasons)?

The key point is that every ingredient added must earn its place in the final mix; the question must always be asked as to what would happen if it were left out?

To explore some of these options, let's take one case study, of a large company producing of an acidic carbonated beverage thinking about introducing a new lemon-flavored variety (a so-called line extension), while in the kind of parallel universe convenient for such thought exercises a producer of organic milk ponders exactly the same possibility.

The first question each might ask is where they might get it, which rapidly changes to a question of what "it" actually is. Where does lemon flavor come from? The suspiciously obvious answer may seem to be lemons, but a different way of answering that might be to say, in lemons, there are a small number of compounds that give them their characteristic flavor, including citric acid, limonene, and a few identifiable others. So, we could consider taking lemons and extracting juice to add to the drink, or else adding a defined cocktail of molecules that give an unarguably lemon flavor. Which would be better? Which would be easier to get, which can be got locally (depending on where the factory was) or needs to be imported (and is that source reliable and secure)? How much will be needed based on projected sales over the next number of years (and how sure are we of these data), and can the supplier match this demand? Critically, how will consumers react to the idea of the product containing a "natural" or "artificial" flavor, and in particular how does this relate to their perception of the original base product and key characteristics and values such as being organic?

Then, the question might arise as to what product is actually in question to become lemonized. The beverage manufacturer might have a low-sugar or full-sugar version and the milk producer a full-fat, low-fat or skimmed variety. Will the perception of lemon flavor be different in these different backgrounds? We can imagine our perception of sweetness or creaminess having an impact on how a fruity flavor like lemon might be detected, and we might need to add more or less of our chosen lemon flavor to achieve the same end result in different bases. Also, if we assume that those consumers who buy or calorie-rich or calorie-light versions are different, which of these groups is more likely to buy our super-new lemony version, as this must influence this decision, and what value will that group place on the natural versus synthetic difference?

In the next step, having identified our target market, our desired base to which lemon will be added, and the material we will add in order to achieve this lemonification, we need to look at our production process and work out how, when, and where to add it.

A key question touched on already is how much to add, as this directly influences costs, and some levels will give an excessively strong flavor while others will be undetectably ineffective, and this will need to be determined, possibly using sensory analysis methods.

The process for making any food product can be described by a flowchart with boxes and arrows, which has raw materials at one end and a finished product at the far end, and the various operations applied arranged in sequence in between (this is basically just arranging the manufacturing instructions graphically). For the new product variant, an edit to this flowchart is needed, with a new step being pasted in, and we need to work out where that is placed. In our beverage, it might simply be one more ingredient added to a big tank of the original formulation, at the determined optimal level, and mixed in. For the milk product, however, it might be more complicated, as the flavor might possibly be changed or its perception altered by a processing step we apply such as pasteurization or homogenization. If the flavor was damaged by pasteurization, for example, we might want to add it after the heating step, but this step is the key way we make sure the product is safe for consumers to drink, and adding something after it risks recontamination if not done properly.

Getting complicated? We are only getting started

Where is this product going to be made? Is it an established plant or are we building a new one? On the one hand, if we are using a plant that makes our main beverage normally, we must be displacing production of that product at least partially to make the new one, and how does that relate to our overall business strategy? If, on the other hand, we decide to set up a new plant, that costs a lot of money, and we need to have real confidence that the product will be a success, that those hypothetical consumers *really* want this product, and that we can predict with high confidence how much of it they will drink in terms of the scale of plant we will need for the future.

Then, having worked out what to add, where and when to add it, and how much to add, we had better check what if anything else it does to the product, besides the contribution of lemon flavor. For example, does it influence shelf life? Does it promote or discourage microbial growth or chemical changes in a way that gives the product a different stability to the original product, and does the flavor itself change during storage? The answers to these questions might be very different for a highly acidic beverage in which the acidity provides a naturally hostile environment for bacterial growth anyway to a more neutral beverage, such as milk.

Finally, after we have worked out formulation and process considerations, we need to remember that this product will have to go into a package, which may need to be created, modified (does the lemon impact in any way on the stability of the package we plan to use?), or at least redesigned (to convey the image and name we have in mind for our modified product).

In the preceding short outline of an example of formulation and process design, a remarkable number of question marks peppered the text, as clear issues that need to be addressed carefully were identified. On the one hand, this was a relatively simple example, though, as for either product base mentioned most of the decisions that needed to be made were already in place, in terms of most of the other ingredients and their supplier and level being well established, as well as most of the processing steps, and only a small number of additional modifications being needed to this well-oiled machine.

Picture, on the other hand, starting with a blank sheet, or a product or recipe concept or idea, for which every one of these pieces of information needs to be worked out from scratch. The complexity is potentially enormous, which is why every analysis of key factors for successful product development will stress the importance of rigorous discipline in the process.

The case of infant formula

One of the most complex materials in nature is milk, from whatever species it originates. This is perhaps not surprising considering the simple fact that milk is arguably the only material we routinely consume that was intended by nature to be consumed! So, it is nutritionally very specific and rich, and every constituent has a reason for being there, which is of enormous biological significance to the infant consuming it.

Obviously, through our lives, we consume predominantly, in many countries, the milk of cows, as mankind has long established the practice of domesticating these particular mammals for the purpose of supplying milk, and from such milk has been industrialized the production of the huge range of possible dairy products.

The clear exception to this common reliance on the milk provided by cows is found during the first year or so of life, when babies and young infants consume their mother's milk, infant formula, or some combination thereof. While cows' milk is suitable for consumption by older humans, it has long been understood that it is not suitable for such a critical stage of life and early development.

If we take samples of milk from a range of mammals, from humans to cows, sheep, seals, and elephants, we will find that they differ in composition, to greater or lesser extents, although the main four broad categories of nutrients are found in all types of milk. These are sugar (lactose), fat, protein (casein and whey protein), and minerals and vitamins.

Milk from almost every mammalian species has been analyzed at some point, and wide variations found to apply. Species differ in the absolute level of each nutrient present in their milk (for example, the percentage of lactose present), and the detailed breakdown of each type of component (such as the fatty acids present in the fat, or the exact proteins that are present or not). For example, milk from polar bears and seals is incredibly high in fat (over 50%), while the milk of rabbits, reindeer, and elephants likewise have a much higher fat content than cows' milk, and reindeer, rabbits, polar bears, and seals have around three times the level of protein in their milk than cows.[4]

Moving to more immediately relevant reference points, the differences between cows' and human milk are less extreme, but, because of their closeness to our hearts (and other vital organs), are much more significant. Specifically, human milk contains around 50% more lactose,[5] far less minerals (around a third the total level of salts), and around half to one-third the level of protein. Very significantly, the exact types of proteins in human milk differ from those in bovine milk, with fewer members of the casein family, less casein overall relative to whey protein, and one of the key whey proteins (beta-lactoglobulin) being completely absent. These are, nutritionally and physiologically, hugely important differences, as the much higher level of minerals in bovine milk, for example, is actually dangerous to very young babies. The study of human milk composition continues to develop our understanding of these components (and the impact of variability between mothers, and factors such as diet, time since birth, gestational age at birth, and many other factors), while advances in component isolation and fractionation mean that formula is being better and better humanized all the time, coming closer to what is unquestionably the gold standard of mother's milk.

So, the challenge in producing an alternative to human milk (for those mothers who cannot, or choose not to, breastfeed their babies, for all or part of the first year of their life), is in scientific and technological terms very significant: to take bovine milk (assuming this is the base), increase its lactose level, get rid of a lot of the minerals (but not necessarily all in the same proportions), reduce the protein level, alter the protein profile, remove one protein entirely, and, while we are at it, change the exact type of fatty acids in the fat globules.

To achieve this, in theory, a "top-down" process could be envisaged whereby milk is taken and subjected to a mind-bogglingly complex sequence of operations to remove, amend, change, or otherwise make these changes. In practice, however, it is (relatively) much simpler to start with raw materials, mostly but not all isolated from milk as separate fractions of interest, and build the product from the bottom up, adding only what is needed, in the exact levels needed, and this is what is done.

So, infant formula today is a formulated food product (probably the most precisely formulated food product in existence, in fact), made to a very defined and scientifically validated recipe. It is also probably the only food recipe for which flavor is not one of the key drivers, at least in terms of the intended consumers being unable to express to us exactly what they think of or how they like the flavor of the recipe.

---

[4] This may lead to be obvious question of "how do you milk a polar bear or an elephant?"; the answer is "very carefully, but for different reasons."

[5] Human milk also contains compounds called oligosaccharides, which are short chains of lactose, glucose, or galactose molecules bonded together, which are found in only trace levels in bovine milk but are thought to be very important for processes such as neural development in infants.

Infant formula design is also a good example of the difference, and synergy, between the disciplines of nutrition and food science. A nutritionist will understand exactly what the newborn infant requires, considering that such formulas will essentially be their only source of nutrients during unquestionably one of the most fragile and sensitive periods of life, and detail down to the last vanishingly small quantity exactly how much of every single vitamin, mineral, fatty acid (for things like brain and nerve development), protein (to supply amino acids, and more benefits besides), and other components need to be present. As a result, infant formula has one of the longest ingredient lists of any food product, as every single thing the baby needs must be present, in exactly the right quantity so that on consumption of the expected daily intake of the formula a baby will develop just as it should, while not suffering any nutritional deficiencies or ill effects.

Nutritionists thus provide the specification, like a list of everything that *must* be present, along perhaps with a list of what *must not* be present, and hands it to food scientists. Their job then is to work out how to make this product, balancing all the various components, and getting them in there are the right levels and in the right form. In addition, almost needless to say, the product has to be absolutely and utterly safe from a microbiological perspective, whether supplied in liquid or powder form, and so heat treatment, packaging, and all other processing steps need to ensure sterility to an extent beyond even the exacting standards applied to other food products.

So, we first obtain the right mix of ingredients, requiring advanced technologies to pluck from bovine milk components such as proteins and lactose that we need, while leaving behind those we don't (such as the high level of minerals present, and possibly some proteins not present in human milk), and add exactly the right blend of vitamins and minerals, as well as probably some fats (vegetable) to supply fatty acids not present in bovine milk. These are then mixed, probably in two streams, one of everything soluble in water and one of everything soluble in oil, which are then made into an emulsion. This then needs to be treated at sufficiently high temperatures to kill everything present, but a complication arises because we know that heat can destroy some of the vitamins that are key components of the formula, so we need to add more of those to begin with, to allow for a certain level of loss during processing. We also know from previous chapters that the properties of proteins change during heating (because of denaturation), and that this is highly sensitive to the presence of minerals such as those we have also added for critical nutritional reasons, so we need to be careful that we don't get precipitation of these components.

Managing the balance of ingredient selection based on nutritional requirements and benefits and potential deleterious effects of heat makes the production of modern infant formula one of the most technologically sophisticated processes undertaken by the food industry, in plants the standard of which is far beyond that encountered for any other product.

The principle is still as for any recipe or formulation though: understand what the consumer wants and needs, design a product or meal that fills these needs as well as possible, and work out a way to deliver this to as high a standard as possible.

16

The Experience of Eating

What happens when we put food in our mouth?

We think perhaps instinctively of our tongues as supersensitive tasting machines, laden with taste buds that detect and analyze core flavors such as sweet, sour, and salty, from which we build up a picture of what the food tastes like.

However, before the food gets there, it has to pass two arguably even more sensitive sensors, the impact of which on what we think the food tastes like is immense. The first is the eyes, and the second is the nose. Controlling the whole system, but perhaps more infallibly than it might think it does, is the brain.[1]

Let's think about the eyes first. We all make automatic judgments about food based on observation, and these first impressions can be incredibly difficult to bypass. Appearance can make complete fools of us, if we let it. I have seen experienced food specialists taste bright orange sweets that comprised apple-flavored jellies with a strong orange dye added and, when asked to describe the flavor, voted verbally for orange, except a few lone and somewhat confused voices claiming for apple.

Famously, even one of the best-known color/character differences in the food world can be hacked by playing with appearances and expectations. Red and white wine can be confused for each other when the taster cannot see the color, and obvious cues such as the temperature at which the wine is tasted are manipulated. This has been demonstrated repeatedly in experiments involving so-called experts in France and the United States, where completely different flavor profiles have been reported on tasting one glass of white wine and one of the exact same wine to which a flavorless red dye had been added. In addition, the flavor profile of wine has been shown to differ depending on factors such as the label placed on the bottle (and apparent perceived "fanciness" as a result), and there has frequently been shown to

---

[1] Of course, some animals make us look like complete amateurs when it comes to the acuity of our sensory systems. My King Charles Spaniel, Juno, can go from completely asleep to sitting with hungry expectation within 10 seconds of a plate of chicken being taken out of the fridge, three rooms away from her bed, while distinguishing and dismissing as undeserving of such enthusiasm the odors of any other meat product under similar circumstances.

be no correlation whatsoever between price on the bottle and the results of sensory evaluation of flavor or desirability.

On the one hand, these kinds of experiments can of course be backed up by highly sophisticated scientific analysis to profile all the flavor molecules and volatile compounds in different wine samples, and molecularly characterize the very compounds that tickle noses and taste buds to build body, bouquet, and other attributes. On the other hand, they can be very simply demonstrated by just putting wine into unmarked glasses, blindfolding the taster, and warming up white or cooling down red wine. The most important agents in the experience of tasting wine really seem to be the eyes and the brain, even before a drop passes the lips.

Of course, between glass and lips comes nose, and indeed much of what we think of as flavor of food is really aroma. We create in our minds much of the final flavor profile we give a food before it hits those expectant taste buds, and a lot of what remains by odors released in the mouth. The term retro-olfaction is used to describe the release of volatile molecules from food when warmed in our mouths, with these aroma-rich molecules taking a shortcut up the back of the mouth into the nasal cavity.

I have seen this relationship between flavor and aroma powerfully demonstrated when a taster was shown a packet of fruit sweets, and then asked to taste one while his eyes were shut and he had his nose pinched shut; as he guessed some random flavor, he had no idea that the person running the test had switched his sweet for a mint before placing it in his mouth, and could not perceive what we might think of as an intensely powerful flavor by mouth alone. His brain was completely fooled, and he could only guess what nonexistent fruit flavor he was supposed to be experiencing.

So, while we often think the terms "taste" and "flavor" mean the same thing, in scientific terms they really do not, as taste best describes the response of very specific sensors in the mouth, while flavor is a global phenomenon derived from a combination of measurements and judgments made in the mouth and the nose (and eyes), and integrated into a global perception and verdict by the brain.

What happens when we then put the food in our mouth is clearly a very complicated phenomenon, or in fact a series of linked phenomena, taking place in very specific orders.

First, there is the sensation of temperature, and whether we regard a food product as hot, warm, cold, or whatever. This can be simple, in that it is correlated with the actual temperature of the food we have tasted, but sometimes it is more complicated, when the change of environment of the food causes a physical change in the food that in itself drives a dynamic change in what we perceive as temperature. To explain this, consider butter, which we might consume cold as a solid freshly removed from the fridge to spread on bread, but that rapidly changes state at body temperature as the fat changes from solid crystals to liquid oil, a transformation that needs energy to fuel it, that energy being drawn from our mouths. So, butter, when it melts, sucks heat from our mouth and tongue and makes it feel cooler than it should be for that temperature. The same phenomenon takes place when we consume ice cream, as the ice crystals suck in energy to fuel their own escape from crystalline bonds into the watery freedom of the molten state.

On putting any food product in our mouth, of course, it doesn't exactly just sit there, but is almost immediately subjected to a wide range of physical forces, as we apply the machinery of our teeth, tongue, and gums to deconstruct the food, chewing, biting, tearing, and compressing it into a form that is more amenable in size and consistency for the journey forth into our lower regions, where it will be stripped for nutritional parts. This process of mastication eventually amalgamates the food in the mouth into a single mass, called a bolus, which is what is swallowed. This consolidation process helps to avoid the risk of random bits of food being inhaled into the windpipe, with potentially hazardous consequences. It is accepted that our mouths also include mechanisms for detecting mechanical properties of food, motion of food within our mouths, temperature as previously mentioned, and pain.

All through this process, initiated by putting the food in our mouths, a reaction with saliva is also taking place. Saliva is a very powerful bodily secretion, which serves to break down, lubricate, and initiate flavor release from food. Chemically, it is a very complex mixture of enzymes, lubricating substances, and also some protective antibacterial substances that help to defend us against unwanted microbial hitchhikers attempting to ride with our food into our bodies for mischievous purposes. Some very significant digestive steps take place immediately on reaction of food with saliva, including the start of the breakdown of starch to simpler sugars, resulting in an in-mouth perception of sweetness even for starchy foods that don't contain much free sugar in their formulation. Saliva also serves to break down emulsions in the consumed food and start to release fatty components, probably helping in the perception of "fattiness." It has been reported that compounds called tannins in wine help stimulate the secretion of saliva, and so might make food easier to eat, which might explain the long-standing association of wine with a meal (not to undermine its many other fine qualities).

The saliva then delivers the products of this initial oral processing to the sites of flavor detection, which may be in the mouth (taste buds) or in the nose (the far more sensitive site). The taste buds on the tongue are the primary sites that process the food-derived information to tell our brains what to think about the flavor of what has just been put in the mouth. It was originally thought that there were four taste buds, for the taste equivalents of primary colors, these being salty, sweet, acidic, and bitter. However, it is now recognized that we also have receptors specifically for a taste sensation called "umami," often described as the savory character most commonly associated with Asian food. It is particularly invoked by the presence of an amino acid called glutamine, or in particular the form of it often found in, or added to, food, called monosodium glutamate. Umami flavor is often described as "meaty" or "brothy," and is said to induce a lip-smacking reaction. Key sources of umami besides meat include mushrooms.

The taste buds are also not independent sensors, but the signals they send can be processed interactively by our brains, with salt or sugar helping to tone down bitterness, for example, while umami flavors attenuate the perception of saltiness and can reduce the level of salt required to achieve a certain response. In fact, salt can neutralize bitter notes in a product, allowing other flavors to shine through, so salt has essentially a broad flavor-enhancing role. That might explain its traditional presence in, or addition to, so many different culinary outputs, even perhaps surprising products like cakes or cookies.

Interestingly, very recently, scientists have reported that we might actually have a previously unrecognized set of receptors for "starch" as a flavor sensation in our mouths. It was previously thought that we tasted such large carbohydrates after they had been broken down by salivary enzymes into smaller sugars, which then tickled our taste buds for sweetness, but these buds can be blocked in volunteers, who can still taste something specific from starchy foods, which suggests that we have separate receptors for larger sugar-based molecules that help us to ascribe a specific taste sensation to foods like bread or rice. Discoveries like this have made scientists wonder whether there might be a much wider range of taste receptors than previous thought, including receptors for specific amino acids, metals, and fatty acids, and certainly acceptance of the idea of "fatty" as a basic taste is growing.

Why do we have the particular set of taste buds that we have? What evolutionary pressures favored these ones? As mentioned previously, this has been suggested to be linked to survival instincts, helping to draw us to sources of energy (sweet or fatty), find foods that help in key physiological processes (salty), or have the raw materials for the proteins our bodies continuously need to produce (umami), while avoiding spoiled (frequently sour) or toxic (often bitter) foods.

Changes in state of the food can also affect flavor release and the timing of our perceiving flavor, as can be seen again taking the example of consuming butter. In cold butter, the salty or milky flavor is trapped in small water droplets immobilized within a crystalline fat prison and is not available to interact with our taste receptors until the prison dissolves, the cell doors fly open, and the water can flow out, those tiny salty droplets coalescing to unleash a wave of saltiness measurable fractions of a second after the melting takes place.

One of the final elements of sensory evaluation being undertaken when we eat that is perhaps underappreciated is the role of sound.

Some examples of very evocative sound effects on eating food include crispy, crunchy, and crackly (the very words themselves give a sense of the sound, don't they?). While we might casually use these terms almost interchangeably when thinking about the texture of a Granny Smith apple or potato chip, for example, they have actually markedly different sensory meanings. This relates to the notes they emit, but not flavor notes for once, more like musical ones. Crunchy foods emit sounds of a low frequency (like bass) while crispy ones reach much higher frequencies, and crackly ones might emit a mixture of the two. We actually detect these sounds and interpret them in our ears by vibrations that are transmitted through the bones in our skulls. When we like eating such foods, we are picking up good vibrations, as the food is clearly sending us excitations (om bop pop). In contrast, the absence of such sound cues can lead us to immediately attribute negative qualities to a food that we expect to be crunchy, like a bag of chips having gone soft.

Finally, we cannot overlook the role of psychology, memory, emotion, and such highly individual and perhaps unquantifiable human factors on the experience of eating. One of the most famous references to memory comes from the French writer Marcel Proust, who in his epic seven-part novel *Á la recherche du temps perdu* (*In Search of Lost Time*) described a profound experience of being vividly brought back to his childhood by the experience of eating a small shell-shaped sponge cake called a Madeleine. There is no doubt that we have all imprinted associations of certain food tastes and aromas with specific experiences, time periods, or important events. For me, this occurs most specifically when certain food smells (particularly roast turkey) invoke a sense of Christmas-ness, or certain fizzy, orangy energy drinks bring me right back to being fed them as a child suffering some minor illness or other.

Emotions such as disgust are also key determinants of responses to certain foods (or even the idea of certain foods, as has been found to be the case when consumers are presented with the idea of eating insects), while some individuals are thought to suffer from what is called food neophobia,[2] or a fear of being adventurous in trying new food experiences.

I have said that our brains are one of our most important sensory organs when it comes to food, but the way in which our vast neural networks so strongly link food to reaction and thought is perhaps one area where food will always be one of the most individually specific manners in which our sensory experience of food will always be unique to us alone.

The science of sensory analysis

How do we make the responses that people have to food something we can measure, analyze, and make decisions and develop products and meals on the basis of?

Sensory analysis is the scientific discipline that deals with trying to systematically measure and evaluate more precisely the way in which we perceive food, as previously discussed, using all the senses (sight, sounds, smell, taste, and touch).

We are, of course, all amateur sensory analysts.

We start very young, when babies make a puckered face on trying something that they don't like, and proceed every time we taste a product and decide whether we like it, and file away in our memories what we like (or not) and why. We also conduct a sensory evaluation every time we subject a friend or family member to something we have prepared or wish to recommend and stand back with a raised eyebrow and an expectant question of "well . . . ?"

---

[2] Most commonly seen in children, but perhaps occasionally also in adults, myself included.

The amateur sensory analyst stops there with a sensation of satisfaction or annoyance, as may be the case. The professional sensory analyst will not stop there though, but will delve much deeper, through a series of questions that might start thus:

Okay, so you like it, but why?
What are the main characteristics of taste/smell/appearance for the product?
Can you rate these on a scale of 1 to 10?
And so on . . . .

Sensory analysis might consist of a group sitting around a kitchen tasting something a chef has prepared and giving verbal or written feedback, or a large group in a laboratory tasting a range of products and entering the results on a computer, providing data that will be crunched (just as the product might also be) into graphs and plots of liking.

Sensory analysis today is a hugely complex discipline, which ambitiously seeks to quantify and understand the most basic and fundamental of human experiences, shared by every living person every day, which is the reaction we have to the food we consume.

To do this, a range of tests and approaches are used by sensory scientists, depending on the kind of information that is required, while the subjects who do the tasting will also differ, depending on the objective to be fulfilled.

For example, in some cases, the tasters whose views we are seeking need to be trained to be sensitive to the qualities being measured, whereas in others there is an advantage to having "naive" tasters (like a panel or real consumers or customers), who give what might be a more common reaction that might be encountered in the real-world population.[3] There are also known to be so-called "super-tasters," who have highly attenuated sensitivity to certain tastes (like saltiness or bitterness) and sense these with a far higher intensity than the average person, while at the opposite end of the sensory sensitivity spectrum lie non-tasters, who have decreased sensitivity and find many food products bland and unexciting. A person's placing on this scale appears to be directly linked to the density of taste buds on his or her tongue, and it has been estimated that a quarter of people on average are either super- or non-tasters, and half lie in the "normal" tasting category.[4]

---

[3] I once heard an eminent sensory scientist at a conference proclaim proudly that his institution had a highly trained panel, populated by individuals who were screened to show that they were in the top 3% of people in their ability to discriminate subtleties in the flavor of the product in question (cheese). An audience member then asked the very reasonable question as to whether knowing about differences between products that 97% of the general population would not notice was useful information.

[4] This seems astonishing to me, meaning as it does that a quarter of all people have essentially much reduced sensitivity to tastes while the same proportion have the opposite reaction. Put another way, look at a typical family of four around a table and ask which one is the over-taster while the other extracts far less sensation from their food. As non-tasters typically compensate for their taste deficiencies by adding more seasoning to their food to make it taste good and have a preference for very sweet and high-fat foods, I know for sure which I am.

Going back to our assembly of a panel to taste our product, if training is deemed desirable, this might consist, in the case where the goal is to evaluate the sweetness of a product, of the tasters first tasting a series of products or even simple sugar solutions to "anchor a scale," so that they all had a common understanding of what might be agreed to be a nonsweet product (say a score of 0 on a scale from 1 to 10) and what would be an extremely sweet product (a score of 10 on the same scale), and could place intermediate levels of sweetness in the right relative order in between. A thus-trained panel would then be able to take the "test samples" and, by comparison with this carefully assembled mental memory scale, place these on the appropriate point of the scale and give what hopefully will be reliable data.

A key point is that, for each food property, we all have different preferences and sensitivities. I have a very sweet tooth, for example, and what I might regard as a perfect level of sweetness in my coffee might sicken another person, while the same might apply to our own consideration of properties like bitterness, saltiness, or sourness.

If we then hypothetically gave the same product to three individuals to taste and rate its sweetness, we might get wildly different views and results. If we then had two products (say what we hoped was a new, improved recipe for a product that we wanted to compare with the original), and we wanted to know whether there was a real difference in perceived sweetness, we would find it very hard to draw any conclusions, because there is a high chance that the differences between the tasters' views will be much larger than the differences between the products.

How can we reduce or control for the differences between individual sensitivities and preferences, so as to be able to more sensitively focus on the analysis of the products and characteristics of interest? One step is training, as just described, but the other key consideration is the number of individuals tasting. The more people we ask to taste and rate the product, the more likely it is that we can get an averaged consensus for the property of interest, by diluting out the more extreme individual scores for those with a naturally very high or very low preference or sensitivity for what we are interested in measuring. So, panels of tasters evaluating products generally get larger and larger the less training we apply to reduce natural variability.

In many ways, a really important part of sensory analysis is about controlling the variables (things that might change the result you get) and reducing these unwanted sources of variability in our data, the better to home in on what we want to measure and understand. What I have previously described is a simple example of how to reduce the variability between the people tasting the product, but many more factors might be controlled in setting up a sensory analysis.

We can often need to control and train, and indeed restrain, the subconscious, by withholding from it the information that we know might just confuse it and mess up our analysis. This could be done, for example, by blindfolding the consumer or, less threateningly, by controlling the lighting. For example, tasting many products under a red light removes visual cues, and color will then be effectively neutralized as an influence on taste, the food equivalent of the curious old adage that all cats are gray in the dark.

Temperature can also be simply manipulated; for example, as mentioned earlier, our perceptions of red and white wine are hugely influenced by the fact that we expect the former to be served significantly warmer than the latter, and serving them at the opposite temperatures to those normally used can very quickly confuse expectations if combined with the removal of the huge color clue.

Of course, we know that the flavor we attribute to a product is actually the sum of the taste we perceive in our mouths and the aroma the far more complex set of sensors we have in our noses detects; increasing temperature has a direct effect on how much of the flavor goes up our nose, as it were,[5] and so the temperature at which a food is tasted has a massive influence on how it will be evaluated, and different flavor properties might be given to exactly the same product simply by virtue of its being tasted at different temperatures. Also, it seems that some receptors responsible for detecting taste in our mouths are very temperature sensitive, so serving foods warmer will broadly increase our perception of the intensity of their flavor.

Indeed, considering how much of flavor is actually judged in the nose rather than the mouth, when we are interested in isolating these effects we might either request our tasters to smell but not taste, or else taste but under conditions where their nasal inputs are negated by simple pinching of their noses. It is amazing how "blind" we become to many strong flavors if our nose is blocked, as we know when we have a cold.

So, before we ask anyone to taste a single bite or drop of our food or drink, we have to carefully set up the analysis, in terms of the lighting, room, and product temperature, number of tasters, and a whole range of different conditions that have been carefully controlled to reduce unwanted influences that will reduce our ability to get good and useful data on what they actually think of the product. This might be done overtly by people in white coats in a laboratory, or more covertly and perhaps even inadvertently by a restauranteur when he or she sets precise conditions for diners on their premises.

Then, we need to consider what questions we plan to ask when our subjects do taste, smell, and look at the product. This all depends on what information we want to obtain.

In a simple test called a *triangle test*, we might just want to know if two products are different to each other or not. This might be our old recipe versus a new, hopefully improved, one, or our product versus that of the competitor, or our product before and after changing an ingredient or processing step. In some cases, the answer we seek might be that they are different, and tasters can distinguish one from the other, while in others we may be happy if they are not able to tell the difference.

If we gave one person two products and asked if they were the same or not, this would probably not give us very good information, as that person might just guess!

---

[5] Increasing temperature increases the mobility and energy of all the molecules present, and those that are prone to adopting a highly mobile gaseous state in particular will be increasingly happy to leap free of the solid or liquid in which they were previously entrapped and more rapidly become part of the aroma of what is being consumed.

To get better discrimination, we would give the person three products (the three points of the triangle in the test name), where the person conducting the test knows which is which but the taster does not (the products might have random number or letter codes as their only identifier) and ask simply which is the odd one out.

If the tasters had no clue but, when pressed for an answer, made a guess, there is a one-in-three chance they might pick the right answer, but two-thirds of the time their answer would be meaningless. If we had two people taste, and both picked it correctly, there is a smaller chance this could have arisen because of guesswork, and with more tasters the chance of getting a correct result by fluke gets smaller and smaller.

A bit of fancy mathematics that we don't need to worry about here (sighs of relief all round!) then leads to tables[6] whereby, for a certain number of people doing the test, we can check the minimum number of those who would have to give the right answer for us to regard the result as reliable.

In scientific terms, we define the word I just used "reliable," very precisely, in terms of what is called statistical significance. Statistical significance is a way in which scientists evaluate the likelihood that something that has just been measured makes them happy or sad, by quantifying the chance that a result or trend could be a random fluke. It is a way of expressing the likelihood that, if we did this experiment or measurement a huge number of times, the result we have just found could come up randomly by pure chance.

To put numbers on any estimation of the likelihood of real effect or difference on the one hand and random fluke causing undeserved excitement on the other, scientists set thresholds for the levels of excitement they are allowed to feel based on the likelihood of a real finding being justified by the data they have.

The lowest level of confidence we might allow in a result that would be regarded as somehow sound or solid is a 95% confidence in that outcome. In other words, there must be at most a 5% chance a result could be achieved by random chance for any statistician to even give that result a passing chance of being noteworthy. If the statistical tests applied indicated that there was 99% confidence in the result (1% chance of it being nature taunting us cruelly) the researcher would be much happier, and the satisfaction would reach its allowable zenith if the numbers spat out a confidence result of 99.9%. In this case, if we ran the test a thousand times, the result seen should come up by chance only once, and so there would be justification for concluding that, yes, this cookie is actually detectably different to that one. A difference or effect that is found to have a level of confidence of at least 95% is described as being statistically significant.[7]

---

[6] The mathematical ones constructed with vertical and horizontal lines and filled with numbers, not wooden ones, that have the vertical and horizontal lines, but are instead filled with food.

[7] It should be noted that because something is statistically significant doesn't mean that it is a good thing or important; it just means that the data can be shown, after analysis of what causes differences between the individual results, that the effect being tested accounts for more of the difference than other factors, including random chance.

Going back to the example of the triangle test for food, there are reference tables that define how many assessors must correctly claim there to be a difference for one of these thresholds of certainty as to there being a real detectable difference to be exceeded. For example, if 15 people tasted two cookies, and 8 or fewer picked the right one as being the odd one out, this could be concluded to be a fluke, or at least not a result to have any level of confidence in. However, if 9 picked the right one, this could be concluded to be the lowest level, 95%, of acceptance confidence in there being a difference; if 10 got it right, we could claim 99% certainty, and if 12 or more got it that would increase to 99.9%. In this way, we can take the data from a simple test and draw conclusions that have differing levels of certainty (or not) as to their validity, and the act of tasting three cookies has now become a powerful test of difference.

This is a basic test applicable to certain food situations, but is obviously quite limited in that it compares pairs or, at best, small numbers of products tested repeatedly against each other in pairs, and gives only an indication of similarity or difference.

Modifications to this basic triangle test that give more flexibility in terms of different kinds of information that can be obtained, again for small numbers of samples, could be to ask the assessors, once they had picked the odd one out (which is all we had asked them to judge so far), which they preferred, and even why, so that gradually more powerful layers of detail can be extracted from this comparison.

A related test might then be a *preference test*, where the panelists are given two samples to taste and told they have to express a view as to which they prefer (which is called forced choice, for obvious reasons), or perhaps more than two and asked to put them in their order of preference. In another variant, called a *duo-trio test*, tasters might get three samples, one of which is identified as a reference, and be asked to say which of the other two are the same as the reference. In all cases, the number of tasters getting the correct answer can be cross-referenced against the total number of testers to determine whether it is safe to conclude that there really was a detectable difference between samples.

What if there were a large number of samples and detail was required on specific attributes, like five different versions of a product with different levels of a particular flavor added, and the goal was to determine the ideal level of flavor to be added? In such cases, we might apply different approaches like *ranking*, where a set of samples are presented to a taster who is asked to put them in order of the characteristic of interest, such as sweetness or bitterness, or even overall liking.

Or we could use *rating*, whereby panelists are presented with a number of samples and asked to rate certain named attributes on a set of scales, such as the range from 1 to 10, as previously mentioned. So, they may be asked to rank a yogurt on the basis of acidity, sweetness, mouthfeel, viscosity, and overall desirability, while different terms would be used for a fruit juice smoothie or piece of steak. The characteristics for any product thus evaluated might relate to flavor, aroma, color, or texture of the product in question. The scores given by different panelists, perhaps replicated by asking them to taste and rate each product for each attribute more than once, can then be statistically analyzed to see which products are genuinely likely to be regarded as having a real difference in that property from the others.[8] What could also be assessed, in one variation on this approach, is the *degree of difference* of samples relative to a standard control comparison sample.

Finally, the most complex type of sensory analysis is probably what is called *descriptive sensory analysis*, where the goal is to create a complete picture of the sensory properties of a product. Usually, the list of what these properties are starts off blank, and the first task of the panelists is to build their own *lexicon* of what they see as the properties of the products to be tasted. They will select a set of the key words and terms, which they can then in a subsequent stage rank the products along. Advanced training can be provided to make sure that the tasters' taste scales are accurately calibrated, with standard samples exhibiting different levels of intensity of the characteristics (like different strength solutions of sugar for sweetness or quinine for bitterness) being used to get the panelists to agree on what represents a very high, low, or intermediate score for that attribute. This method will give the most complete picture of the sensory characteristics of a product, but inevitably is the slowest and most expensive method, requiring lots of time and effort to get the tasters to the point where they are unleashed on the products with their hypersensitive descriptive powers.

So, we can see how a progressively more complex set of questions can be addressed to put real science behind how people taste and respond to food, while applying rigorous mathematical analysis to the data to make sure that wishful thinking doesn't lead the tasters to imagine they see something in their data that isn't there and that real conclusions can reliably be drawn about the data.[9]

Sensory analysis might be also used to test a product and how it changes over time after production. The flavor of some food products, as discussed throughout this book, evolves over time, and the food products develop different and more complex flavors, sometimes getting progressively better with age, like wine or whiskey, and sometimes they reach a peak of quality from which they go downward, like some types of cheese. In these cases, sensory analysis can track the changes in flavor over time and help to understand the changes and consumer appeal, and ideally be

---

[8] In the statistical test used in this case, called analysis of variance (ANOVA, pronounced as it reads), the question is whether the variation in scores between products is greater than the variation between the scores given for the same product by multiple tasters or by the same taster when testing the same product more than once. If the variation between scores awarded by tasters is much more than the difference between the results for two samples, it will be concluded that there is no significant difference between the samples. Sensory analysts use a lot of statistics, often over ANOVA.

[9] One of the most famous basic statistical tests, the Student's t-test, was developed in 1908 for the specific purpose of comparing the quality of different batches of raw materials for the brewing of Guinness. Things have clearly changed in 100 years, as no Irish student would now ever confuse t and Guinness. Interestingly, more broadly, the field of agriculture has also been a massive influence on the development of statistical methodologies, and terms applied in all kinds of fields (pun sadly inescapable) such as "split-plot design" owe their names to designs of experiments to compare treatments for the growth of crops.

linked to chemical changes in the product, so that ultimately the relationship between molecules and meals can be unraveled just a little bit more.

The future of sensory analysis?

All of what we have previously discussed involves people tasting food and giving
their views in the most controlled and carefully planned out manner possible.
However, humans are quite imperfect tasting machines. People, as mentioned
earlier, have different preferences and levels of sensitivity to different flavors, but
also have moods and illnesses and sometimes hangovers, as well as different diets
and likelihood of having tasted something else before the taste panel. Some smoke
and some don't, with major consequences for their taste sensitivities, and there are
a million other confounding factors that make people different to each other, and
they even differ in their own behavior on different days or at different times of the
same day.

Taking all this into account, wouldn't sensory analysis be much easier if we
could remove people from the process entirely? What if we had machines to do it,
that would show up every day and behave the exact same every time, and not need
payment, thanks, or a treat afterward?

If tasting food is really the sum of a series of chemical reactions between
molecules and receptors or sensors in the body, can we not design a machine that
captures and measures those reactions?

For all these reasons, there has been significant interest in recent years in the
development of analytical methods to supplement, and perhaps eventually replace,
human sensory analysis. Any modern conference on sensory analysis will include
talks on such cybernetic-sounding topics as electronic noses and electronic tongues.
As the names suggest, these are instruments that are designed to replace the equivalent human sensory organs and provide data that can be used as surrogates to
predict what an actual (somehow typical) person might respond to the product
under analysis.

Assessment of color can be made into an instrument quite readily using an
instrument called a colorimeter, which measures the color of a product under
controlled conditions of lighting and presentation, using what is called a Hunter
LAB scale. On this scale, three main values are generated to describe the color
of a food sample; the L-value measures lightness on a scale from 100 (white) to 0
(black), the a-value measures color on a scale that goes from red to green, and the
b-value places samples onto a hypothetical line from blue to yellow. Taking these
three numbers, and some other parameters calculated from them, allows an objective color profile for a sample to be determined, and then changes between samples
(say color changes that are due to cooking meat, or browning of an apple) can be
expressed in very precise numbers.

Systems have also been developed in which texture can be measured using
instruments called texture profile analyzers, which basically subject a sample held

on a metal plate or in a cup to some sort of force by a probe approaching from above, and variously puncturing, cutting, stretching, or twisting the food (depending on the probe chosen and its mode of motion) while measuring very sensitively the resistance of the sample to the applied action, which is stronger the more the food is firm, stretchy, or viscous. From such tests, properties like firmness, gel strength, viscosity, cohesiveness, springiness, and many more can be expressed in numbers, and again differences between samples or variations over time during storage can be examined without the need for an actual person to pick up the food and chew it, squeeze it, or apply any other physical force we routinely apply to food.

Some instrumental systems hook up people and machines, as when a sample of aroma compounds from a food product are separated based on properties like their size (using techniques such as chromatography, as described in the Appendix, but operated in a stream of gas) and, when each compound in a mixture passes a detector that records its presence and some identifying property, like the weight of the molecule, then it exits and passes without further ado along a tube and up the nose of a tester (person) who records observations on the aroma of that molecule, so that compounds 1, 2, and 3 in the mixture, which are likely to equate to molecules X, Y, and Z, smell "sweet," "chocolate," and "bitter," respectively.

There are also in development or launched as prototypes apps that can take a picture of a dish of food and, by comparison with vast databases of images, provide estimates of the ingredients, composition, and every calorific content of a food product.

Of course, given the fact as discussed throughout this chapter that psychology plays a key role in how people perceive food, can we ever have machines with the attributes to fully appreciate all the properties of food a person thinks about when they see, smell, and taste a real product?[10] The answer is presumably, and desirably, "no," but there is probably no doubt that such approaches will increasingly complement human observation in routine quality control to ensure consistency of production and perhaps even safety.

---

[10] Probably not even Skynet.

17

The Kitchen and the Lab

Just a matter of scale?

To this point, my focus has been largely on the transformations and processes that convert raw materials and ingredients into packaged final food products, while considering the relationships between such a scale and what happens in the kitchen. I believe that food science is food science whether it happens on a 10-tonne scale in a factory or in a kitchen at home or in a restaurant. It is just a matter of scale.

But is this really a defensible proposition?

As pointed out several times already, all food products consist of a set of raw materials and ingredients, which we submit to a process or series of processes and then place in a package in which it should remain safe and suitable for consumption for a defined period of time.

What about a meal? Ingredients and raw materials, check, just taken from a larder, fridge, or freezer. Processes? Check, just maybe a different set and scale, as will be discussed.

Package and storage? No, but one could say the plate, room, atmosphere, and a million other elements of presentation of a dish at home or in a restaurant are the package. Likewise, being able to maintain a shelf life may not be a priority, but it is usually regarded as a good thing when safety for the eater is guaranteed, while we often hope that those leftovers we put in the fridge or bring home in our doggy bags will retain some form of safe edibility for at least a while.

Food science is science above all else, whatever scale it happens on. In the kitchen, our raw materials, animal or vegetable in particular, are products of biology, while the reactions that take place on plate or during cooking (or other processing steps) are driven by chemistry, and physics determines what happens when we heat, cool, mix, or all the other things we do. Like I said in the Introduction to this book, even the humblest kitchen is a highly scientific environment, and every meal is an experiment.

Processing, chemistry, and the kitchen

I have discussed the big building blocks around which food processes are built, such as heating, cooling, freezing, and drying. Each of these clearly has it place in the kitchen, and we have already seen how the principles of heat transfer apply equally well in a frying pan as in a pasteurizer. Chefs might not operate spray-dryers, but reduction and concentration by heat are very familiar, and the effects of pressure on water removal are exploited every time we put a lid on a saucepan or use a pressure cooker, while evaporative cooling as water boils off is likewise a familiar phenomenon.

The goals of processing are likewise not dissimilar, being first and foremost about safety, as when the barbecue chef makes sure burgers are cooked through, using heat-transfer principles implicitly, to make sure the heat reached in the dead center is sufficient to ensure that this use of the term "dead" is the only one to apply. We obviously don't, in such circumstances, monitor the heat load by counting microbiological survivors or assaying enzymes that tell us if the required temperature was reached. We do use proxies nonetheless in terms of clues such as juices running clear, which tells us instead about protein denaturation (at the required temperatures) that has stopped the colorful proteins from being easily exudable. Such approaches, science based at their core even if practiced long before the science was explained, are universal across the world of cooking. This is just as was shown in Chapter 11 when the complex thermodynamics taking place in a heated pot, with or without stirring with a wooden or metal spoon, were considered.

Of course, when we cook, the most visible manifestations of progressive change are all thanks to food chemistry. The role of denaturation of proteins in transforming the structure of meat, particularly if combined with even modest heat-induced water removal, has already been discussed, while the presence of many food sugars (naturally or added) give us browning through the Maillard reactions, in foods from steaks to cakes. In many cases, fats also change their properties at high temperatures, melting, flowing, and sometimes transforming, or we rely on their liquid–solid transitions at low temperatures to control texture.

There are few components of the uncooked mix that don't undergo some kind of chemically based transformation when we cook. Some products of such reactions do not stay within the food, but even these are important, as those components that are naturally present or are produced by cooking that can escape into the air as volatiles contribute to our cooking smells, which undoubtedly interact with our psychologies, expectations, and taste sensations in complex ways.

We instinctively apply heat-exchange principles in myriad other ways, like when we place things on racks to cool that expose their maximum surface area to cool, circulating air or know that the top of the fridge is likely to be warmer than the lower regions. We also often witness the dangers of freezing for food quality when we find exuded liquid in our defrosted packages of meat or fish, or ice cream left and forgotten until it has turned into a frostbitten icy mess.

We exploit chemistry and physics casually when we add salt to water before boiling and when we marinade to change the properties of proteins by changing their environment, in particular the acidity. Acids such as citric acid from lemon juice and acetic acid in vinegar, when present in marinades, disrupt bonds in proteins and loosen up their structures, while the presence of alcohol in marinades (like when we add wine) helps to dissolve fats, making it easier for the acids present to get to proteins and soften tougher meat. We have also seen earlier in the book how enzymes can break down proteins in a whole range of food products (such as cheese), and adding enzyme-rich ingredients, like ginger or pineapple, also softens meat by literally predigesting it before cooking and consumption. Again, chefs have optimized their marinades by trial and error for centuries, but these are no secret sauces with magical powers, as every single effect can be understood in terms of protein chemistry.

So, throughout the gamut of culinary activities, we exploit scientific properties of food constituents routinely. For example, as discussed in Chapter 3, whey from cheesemaking contains a very low level of protein, but technologies have been developed to effectively filter out and recover this protein. We can then exploit the fact that it has really interesting properties, particularly when heated, that result in all kinds of textures and structures, like emulsions, gels, films, and thickened sauces. These applications are of interest as much to the chef as to the food processor. If we take a container of whey protein dissolved in water and heat it at cooking temperatures for just a few minutes, it gels into a solid that has measurable physical strength. Today, a very wide range of labels shows that products from sauces to soups include whey-derived ingredients as an ingredient precisely because of properties like this, while chefs use it as an egg white replacer for new types of products, and can even make fibers, films, and lots more from whey proteins.

Food processing in the kitchen: Some unique tools

We have discussed several tools that are used in the kitchen to exert specific scientific effects, such as the microwave oven, frying pan, and oven. Obviously other commonplace tools in the kitchen such as scales, measuring cups, thermometers, timers, mixers, and strainers all have equivalents in a lab or processing environment, as measuring (whether temperature, weight or volume) is a basic operation in all cases.

There are other key elements of the toolkit of the chef, however, that are more unique to processing on this small scale, and the first of these is the knife.

Even the knife has analogues in the broader world of processing, as we think not so much of its appearance, shape, and sharpness, but of what it does, which is to reduce the size of things and thereby increase their surface area, rendering them more efficient or more convenient to cook, consume, or both. We also use it to separate elements within a food, whether to be cooked separately or to cook on the one hand and discard or use for different purposes on the other. In many processes discussed so far, a key goal (for example, when heating or drying) was to increase the surface area over which an effect of interest could then take place, and so it is with the knife. In addition, separation, whether by distillation, membrane filtration, or size classification, is one of the core processing principles of the food industry, into which the dissective abilities of the knife fit quite neatly, as if into a smooth wooden block.

The function of the knife is to cut, for convenience, preparation, or separation, and the key indicator of its success in achieving this, its most fundamental function, is the sharpness, and thereby the precision and neatness of its cuts. With any knife, the key principle of physics that underpins its action is the transfer of a force applied in one place to another place with amplification by focusing it on a much smaller area. If we take a knife and force its point downward into a piece of meat, we concentrate all the force we apply at the handle end and make it act at a very small area, literally a point. This is the same as when we apply force using a hammer to a broad end of a nail, and that force then travels along the nail to the point, to force the nail into the wall, wood, or other surface against which it has been placed.

If we instead use the knife to chop, we apply force downward, again via the handle, which is expressed across the narrow surface that is the cutting edge of the knife. The sharpness is essentially a measure of how small that surface is; two knives of the same apparent outward dimensions, one much newer than the other, can have very different cutting abilities that can be explained very readily by a microscopic examination of the edge dimensions or measurement using fine calipers. The knife with the thinnest edge will transfer our applied force more faithfully and cut more keenly, while a heavier knife will, by virtue of its weight, put more force behind the cut and so give a sharper separation than a lighter one. Further examination of our hypothetical new and old knives might perhaps show unevenness in the older knife, in that the edge is not straight, but pitted and ragged, so that force is not applied in a single direction, but diffused in chaotically different directions at the edge, also giving less even and efficient cuts (unlike a serrated edge, where force will be applied in more than one direction, but in a very controlled way, and where cutting the food in more than one direction also helps to achieve the optimal desired effect). An older knife is also likely to have a more rounded edge, so that the force applied is also diffused and spread out as a result.

As well as chopping, we might use the knife to slice, moving horizontally across the material to be cut as well as down. The difference in whether a material is easier to chop (like a raw carrot) or slice (like meat) depends on whether the material is easy to create a fracture in, whereby the action of the knife creates a tear that will extend easily through the material. The tendency of the materials in the food to split apart under force (and in turn their relative attraction for each other) determines whether the material cuts easily with a chop or we need the extra force caused by the horizontal (or near-horizontal) movement of a slice to get it to pull apart.

But what exactly is cutting (chopping or slicing) doing to the food? To truly consider this, we need to turn our microscope not on the knife surface, but rather on the food with which it has been in contact.

Many of the foods to which we will apply a knife have a cellular or other complex structure (perhaps held together by cell walls, or the intricate networks we have seen that proteins and polysaccharides can build up). Often this is what gives our food its texture or plays other roles such as keeping separate things that, if mixed, might change the quality or character of the food. By applying a knife, we accept that some of these links and walls will be breached, and in general it is preferable to do so in as limited a way as possible. An analogy might be getting through a window by cutting out a very precise circle with a sharp cutter, rather than by smashing it with a rock.

A sharp knife will allow precise control of the damage caused to our food; a less sharp one can lead to chaos. The French food scientist Hervé This famously demonstrated this principle by placing cut onions in a magnetic resonance imaging (MRI) scanner, more normally used for diagnostic medical investigations, and showing that the high level of cellular damage caused by a dull knife would result in the increased release of components that could undergo oxidation, thereby negatively affecting the flavor of the onion.

Another key process that we rely on in the kitchen, perhaps even more visibly than in a conventional food processing operation, is the mixing of ingredients, often the first step in preparing a meal or dish. Most dishes involve multiple ingredients, and before baking, cooking, or otherwise transforming these they must be mixed into a homogenous mass, be it a dough, sauce, or pie, and whether we use a spoon, mechanical blender, Kenwood Chef, or Thermomix, a food processor that combines a huge range of kitchen operations into a single unit.

If mixing solids into water, we immediately appreciate the hydrophobic properties of dry materials, as well as their ability to imbibe water and disperse readily therein. If the material clumps, settles, or needs hot water to dissolve, we are recognizing the differences in reaction of different materials to water, while the importance of energy in the system can be seen, from the helping hand heat supplies to the benefits of more rather than less vigorous mixing. Whether mixing or dissolving, as for so many aspects of food processing, surface area is a key driver of speed of achievement of the desired end result, and the more we can divide up the entities to be brought into contact with each other, the better.

A final item of kitchen equipment is a chopping board, or more likely a range of these for different purposes, made from wood or (possibly color-coded) plastic. Besides providing a smooth and stable surface for operations like buttering bread or cutting cooked foods, one of the key scientific principles involved in the design of such apparently basic items is actually food safety. Good kitchen practice will keep boards used for raw meat separate from those used for any other purpose, to prevent any risk of cross-contamination, while cleanability is obviously going to be a key criterion for preventing the lingering or adherence of tough bacteria. These can lurk on surfaces or in microscopic cracks or pits therein, growing or perhaps just biding their time until they can infect the next batch of material placed on the board. Plastic boards are easy to sterilize, for example, in a dishwasher, but can be more easily scored and damaged (leaving zones for microbial infestation to gather) while, for reasons yet to be fully explained, bacteria can last longer on plastic than wood, perhaps because of some inherently microbially hostile property of wood.

More adventurous kitchens today (such as those practicing molecular cooking, modernist cooking, or such practices, which are explored more in the following discussion in this chapter) might also contain other pieces of equipment more familiar to a conventional laboratory setting. For example, filter funnels or filtration systems that use a vacuum to add oomph to their separations might be used to clarify stocks or juices, while the other main physical process for separating things in an industrial setting, the centrifuge (which we last met being used to separate cream from milk), might be used to separate emulsions, juices, extracts, and lots more besides.

Even processes such as fermentation are being increasingly used by chefs or at home, typically to allow certain materials (such as vegetables or juices, perhaps with added spices or salt) to change and transform because of the unseen metabolic actions of naturally present bacteria or yeasts, or sometimes purchased cultures, giving products such as homemade sauerkraut, kombucha, and yogurt.

Molecular gastronomy and note-by-note cooking

It is probably fair to say that one of the most notable revolutions in the world of food in the last few decades has been the emergence of what has been called molecular gastronomy.

This term emerged in the 1980s as an outcome of a unique collaboration between chefs and scientists, in particular the French chemist Hervé This and the late Hungarian physicist Nicholas Kurti. The original objective was to underpin the understanding and eventually control of phenomena happening in the kitchen with rigorous scientific understanding. This led, for example, to the creation of an ever-expanding catalogue of "culinary precisions," which sought to formalize traditional kitchen practices into more defined procedures, and in the process subjected many of the traditions and legends of cooking to formalized scientific investigation. One of the inspiring ideas for this project was said to be the fact that we understood better the physics of what went on in the atmosphere of Venus than the reactions taking place during the baking of a soufflé.

Looking back, the idea that it might be possible to take essentially the entire world of cooking (or, at least initially, the world of French cooking) and break it down into a list of operations and tasks that could be first formally defined, then evaluated for their logic or rationale, and somehow discard those that didn't seem to have a definable basis for their use, seems incredibly ambitious, yet year by year the project continued toward just this goal. Topics such as whether certain foods could be made by only certain individuals at certain times of the month were considered, alongside the influence of the metal from which kitchen utensils were made on the properties of dishes created through their use.

As part of this process, a mathematical rigor was brought to bear on the way in which food is described, leading to the emergence of new descriptive practices, and food being characterized by arcane-looking equations very far from those one might expect to see in one's cookbook.

For example, as discussed earlier in this book, an emulsion, such as we find in many dairy products, mayonnaise, and sauces, may consist of oil dispersed in water (like cream), or water dispersed in oil (like butter).

The structure of the former may be presented in a shorthand like this:

O/W   Oil (O) is dispersed in ("in" being represented by the symbol /) water (W).

Butter is then, by the same logic, an inversion of the emulsion in cream:

W/O

Also as discussed earlier in the book, many foods, from a mousse to the foamy head on a beer glass, are foams. These products contain a gas, represented by the letter G.

So, a foam might be

G/W   where the gas bubbles are surrounded by a liquid.

While whipped cream, which contains an emulsion of milk fat in an aqueous mix, with air whipped in, might look like this:

G/(O/W)   gas in an (oil in water emulsion).

Getting confusing? We are only getting started, as we haven't added solids (S) yet, which might be present, for example, as cocoa pieces in a milkshake, which might give this:

(O + S)/W   oil and solids, both dispersed in water.

And so forth, up to the magnificent complexity of Chantilly sauce, the formal description of which looks like this:

(O + (W/S)/W)/W.

What do these expressions tell us? We are used to seeing a food product or recipe described by certain characteristics, specifically a list of the ingredients that went into them and the process to which we subjected them, such as preparing, measuring, blending, cooking, and so forth, and together these define the recipe. We also describe a product or meal using words to capture what we see, taste, and smell.

None of these, however, tell us about how the product is constructed and how the various components we have put together have locked into their roles in structure, let alone how that structure is built. The preceding expressions are like the details you get when you want to go beyond looking at the outside of a house or a piece of machinery, and want to see the detailed schematics; they are the architectural plans of food. If food is assembled like Lego from the building blocks of food

molecules, like proteins, fats, and sugars, then the previous odd-looking expressions
are the colorful booklet that shows how the blocks are put together, step by step.

Does this matter to the people eating the food? Do they need to see this on their
menu or food label, and would it any way enrich their experience? The answer to
these questions is clearly no.

Whom they matter to are those responsible for creating the product, in factory
or kitchen, and trying to do so as well and as reproducibly as possible. Above all
else, they provide a toolkit for those interested in innovation to see what could be
invented that has a form and description that has not yet been tried out.

We could, for example, picture a long list of "tried-and-tested" expressions
and the foods they represent, compare this to a theoretical list of all the possible
combinations of O, W, G, and S, plus their "operators" like "/" (and others not
mentioned here for simplicity). It can rapidly be seen that only a fraction have been
turned into food products, and the map of all the possible structures and variations
thereof still has large areas labeled only with the culinary equivalent of "here be
dragons."

There are also databases that list huge numbers of chemicals that can be detected
by the human nose, alongside their perceived flavor or aroma, which can be used as
the basis for designing flavor profiles of products, including designing whole new
combinations and concepts.

This leads to a concept borrowed from another branch of science, called "combinatorial chemistry," which involves considering all the possible combinations of
chemicals and reactions that could be put together, with a view to producing lots
and lots of outcomes and products, which might lead to a new drug, for example.

How does this relate to food? Consider the space shuttle, or the International
Space Station. These are, for the duration of time over which a particular team
of astronauts call them home, closed systems, with limited ability to add new
ingredients (no shopping!). Everything that is needed for sustenance must be sent
with the crew at the start, as must what is inevitably a relatively limited set of
food preparation equipment and operations. So, there are a limited set of food
ingredients and raw materials (further limited by considerations such as the need to
be absolutely safe, reasonably stable, providing a certain set of nutritional benefits
to keep astronauts healthy, and mitigating against things such as bone weakening
caused by prolonged periods of weightlessness), and a limited set of what we can
do with them. How can we combine these in as many different ways as possible so
as to yield variety and positive eating experiences? That is where combinatorial
approaches and systematic food design can come into play, and shows why today's
space travelers are in some senses gastronauts, just with a silent (or zero) "g."

The conceptual framework around molecular gastronomy could be considered
to stretch in two different directions: the future and the past. The first is more associated with the highly exclusive and expensive restaurants (such as the Fat Duck in
England or the Cellar de Can Roca in Spain), which serve what is essentially (and
this is not meant in a bad way) science fiction food, developed and produced in
facilities that look more like laboratories than kitchens. However, the principle that

all recipes are scientific formulations, that all operations in the kitchen are scientifically describable, and that cooking is at its heart controlled chemistry (with side dishes of physics and microbiology) applies to even the most traditional cooking practiced for centuries or longer.

So, the quest to categorize and define kitchen practice by scientific terms and principles must first reach back into the past, and collate, catalogue, critique, and ultimately characterize everything from boiling an egg to assembling a complex pastry, sauce, or entire meal. Then, having laid down the key concepts and rules, the discipline can pivot from describing and explaining what we know, to building on that knowledge to design and invent that which does not yet exist. The process thus goes from being descriptive and ad hoc, as in the past, to being rigorous and defined, while also predictive and able to design new dishes and products.

The term "molecular gastronomy" always for me resonates with the term "molecular biology," perhaps deliberately, but I think this is actually a perfect parallel. In the 1950s, biology went from being a very observational science, which gave a good description of the characteristics of organisms and the principles of heredity by which these characteristics were passed between generations, to a new paradigm where the basis by which this happened, at the level of molecules, was fully understood. The key actor in the process was identified by James Watson and Francis Crick at the University of Cambridge as being DNA, and unraveling the relationship between the chemistry of this complex molecule and the storage and passage of genetic information opened a whole new window onto the world, with direct implications for our understanding of medicine, disease, evolution, and far more besides.

I am not making a claim that molecular gastronomy will have an equivalent impact on the world as we know it, but the analogy holds, in that the newer discipline likewise seeks to go below the surface of observing characteristics of food products and the transformations we induce in the kitchen to detailing the underpinning mechanisms in terms of the molecules and scientific mechanisms responsible. We can then, through such understanding, go from a position of knowing what happens to knowing how and why it happens, and then onward to being able to control and manipulate these phenomena for new ends and purposes.[1]

Note-by-note cooking

Today, the original framework for molecular gastronomy has spun off lots of new branches and offshoots. For example, "note-by-note cooking" proposes starting not with traditional raw food materials at all, but with purified compounds or

---

[1] It just shouldn't go *too* far. One of my favourite pieces of food humor was when the wonderful TV show *Parks and Recreation* brought its main characters to a trendy bar practicing drinks mixology, offering such treats as vodka in the form of a flash of light, whiskey applied as a hand lotion, and a martini in the form of an aromasphere that needed to be attacked with a chisel before consuming it.

molecules. The molecules used are those known to confer texture and structure, like proteins and polysaccharides, and those known to confer certain flavors and colors, whether individually or in carefully selected combinations. These are then assembled in prescribed formulations that either replicate the characteristics of a conventional product, or, in much more innovative applications, create products that cannot be created starting from the traditional point.

The term "note-by-note" was chosen to reflect the analogy with a complex piece of music, such as might be produced by an orchestra; even the most multi-instrumental symphony can still be represented by a series of individual notes. In this analogy, the traditional food ingredients might be the instruments, but the molecules are the notes, and the music can be built up not from instruments alone, but from the more fundamental building blocks that are the notes.

As is a recurring theme in this book, every dish or food product is a combination of molecules and could be stripped back to express them in these terms. Take a sandwich with meat, butter, cheese, and bread; the constituent molecules in each can be listed, all the proteins, individual fats, carbohydrates, and more besides listed, as if we had put the sandwich into a blender and transformed it into a homogeneous smoothie-like paste (don't try this at home!), which we had then fed through machines such as mass spectrometers to produce a full laundry list of compounds present. There are likely to be hundreds, if not thousands, of compounds, in our sandwich, but of these there might be a very small number that really dominate the product's characteristics, either just by proportional weight present or in terms of providing the main properties of texture, aroma, and flavor.

So, having thus identified the key compounds that define our sandwich, such as the starch and proteins that make up the bulk of structure, for example, we could in theory purchase these in purified form from suppliers of food ingredients or chemicals and lay out a series of bottles or jars containing the individual compounds. These could then be mixed, manipulated, and processed in suitable ways to somehow assemble a sandwich, or something approximating it, but starting from a very different point than if we had started with packets of bread, ham, cheese, and a block of butter.

I don't think note-by-note cooking has attempted to make a sandwich from first principles, as it were, but many of the demonstrated examples are far more complex. Key approaches include identifying and controlling food structure, in ways illustrated throughout this book, and in particular creating an alchemist's apothecary of compounds that generate specific food flavors and aromas.

Beverages are perhaps easiest to imagine being prepared by this approach, as the required compounds can be simply mixed with water, perhaps with a dash of ethanol for a kick, citric acid for a fruity note, glucose or fructose for sweetness, beta-carotene or anthocyanins for color, and amino acids, phenols, or a host of other compounds for flavor. Imagine a bartender in a fancy bar making a cocktail not from a range of bottles of exotic blue, orange, and colorless liquids shaken stylishly in a metal jar but from a larger range of small bottles and jars of liquids and powders—would this seem that different?

Making solid foods from first principles is intuitively more complex than liquids. Although one of the themes of this book has been the ability of molecules such as sugars, fats, and proteins to build, control, and transform the huge array of textures and structures we see in food, both raw and cooked, engineering them from pure compounds to replicate convincingly exactly what nature and evolution, plus in many cases centuries of careful breeding, have created is a challenge not to be underrated. There are even different grades of note-by-note approaches, from the toughest challenge of using pure compounds only to more practical variants where extracts, essences, and mixtures can be used as building blocks; the difference between approaches has been likened to the difference between making music on a grand piano or on a cheap children's keyboard.

The first note-by-note meal was prepared by Michelin-starred chef Pierre Gagnaire and served in Hong Kong in 2009; the first dish included jelly pearls that tasted like apple, a crunchy, lemony sorbet, and a caramel strip, and its preparation notably involved exactly zero use of apple, lemon, or caramel. Instead, individual compounds that give rise to such flavors generated exactly the right taste response on reaction with the taste buds, expectations, memories, and psychology of the diners. Indeed, one of the key disciplines involved in note-by-note approaches is neuroscience, as stimulation of the trigeminal nerve is key to sensations such as heat (such as from a chili), cold (from mint), fizziness, sweetness, and sourness, and compounds that evoke such sensations can be key to ensuring the right reaction on tasting a product.[2,3]

Note-by-note principles can even be applied to the improvement or upgrading of existing food products, such as making a cheap whiskey taste more expensive by the highly economical expedient of adding a few drops of vanillin, based on studies that showed that mature expensive whiskies typically contained higher levels of this compound.

Much work continues on understanding exactly what compounds contribute to flavors of different foods, which is key to the success of this approach, and criticisms of early examples of the art that flavors did not fully resemble the "real thing" are being addressed by ongoing studies of the way in which the more dramatic and

---

[2] As mentioned in Chapter 2, the key compound in chilis (and peppers), capsaicin, triggers the burning heat of food containing the compound by stimulating the trigeminal nerve, which is one of the key pain pathways in the mouth. Eating a hot curry literally is the oral equivalent of self-inflicting pain in the body as surely as, but less pleasant than, hitting yourself in the head with something hard. The body's response to pain involves the release of natural painkillers called endorphins, which generate a wider sense of pleasure, so pain is followed by relief, and we are basically manipulating our own body chemistry when we eat these foods, even perhaps (it has been suggested) in a potentially addictive manner. Interestingly, capsaicin is also soluble in fat but not water, which is why drinking water doesn't necessarily put out the kind of fire hot food can ignite in your mouth, but anything with fat in it—even milk—can do a better job. So, milk can be the cure if it gets hot, hot, hot, and leave your taste buds feeling just like heaven.

[3] Another profound illustration of the relationship between neuroscience and food flavor is that the onset of neurogenerative diseases such as Parkinson's Disease can often be preceded by a condition called hyposmia, whereby a person's ability to detect odors, such as from food, diminishes.

obvious "top" notes in a dish that might represent the main characteristic flavor need to be balanced by much more subtle notes, the presence of which might not even be obvious but the absence of which leaves a notable (no pun intended) and perhaps even identifiable gap.

One of the ideas behind note-by-note is reducing food waste and spoilage, as well as costs of transportation of water (as we know to be both the most abundant and most challenging ingredient of most food products), by preparing the food when it is needed where it is needed. However, it remains likely to be a specialized, if very interesting, activity for the foreseeable future, and from a scientific viewpoint represents once again a really interesting frontier where the worlds of the chef and those of the food scientist have merged into direct constructive collaboration, and the fruits or such activities may have outcomes and benefits as yet hard to imagine. In this spirit, every year, in Paris, an international competition is held for students to show off their latest creations on these principles to keep stimulating the spirit of gastronomic experimentation among the chefs of the future.

Combining the science of food and the art of cooking

In the final series of images in this book (Figures 17.1–17.9) I have included a series of beautiful pictures of dishes produced (both food and picture) by my contributor, Brian French, which collectively illustrate very nicely many of the principles of science explored throughout this book, and so in a way sum up the themes I have tried to explore.

My first chapters explored the constituents of food, from proteins and fats to lipids and water itself, and how they underpin the characteristics of every food product. I then discussed how these basic constituents are transformed by a set of physical processes that are similar in their nature whether applied on a large scale in a plant or on a small scale in the kitchen.

These transformations frequently seek to make our food safe to eat (whether we are pasteurizing or cooking) and, in the process, change the very nature of the food itself, through a range of complex chemical phenomena. One of the most common of these is the Maillard reaction between certain sugars and proteins, with its wide array of resulting flavor, aroma, and color changes. While this might be most readily associated with the browning of meat (like the pork in Figure 17.3 or the ribs in Figure 17.9), it can also be seen in meat-free contexts like the whole roasted celeriac shown in Figure 17.2, where the sugars from the celeriac and the honey baste have reacted with proteins to give a nutty aroma and flavor, not to mention a deceptively steak-like outer appearance (at least in this photo).

Other processes we apply, whatever the scale, are designed to separate components of our food, and are often applied in such a way to do so as gently as possible. In the case of the tomato soup in Figure 17.1, ice filtration has been used to clarify the soup. This method involves freezing a thick broth or soup containing gelatin (added, or present in meat that is part of the recipe). At freezing temperatures, gelatin forms a thick network, as the removal of water into ice progressively concentrates the gelatin into a smaller volume, until the molecules completely intermesh and form a dense mat of entangled molecules as they simply can't get out of each other's way. This gelatin filter then entraps particulates and lumps present in the mixture and, if this frozen mixture is then left for a while in a fridge atop a strainer, the ice melts and carries with it small soluble sugars and flavor-rich molecules, which drip through as the lovely clear soup.

Gelatin is a protein, and changes in proteins are key to many cooking-induced changes, involving the unfolding of complex three-dimensional structures followed by the formation of new structures, as amino acids interact with new partners or other molecules present in the food, such as through the Maillard reactions seen in the celeriac example. These reactions are incredibly sensitive in terms of their consequences to a multitude of factors, such as the temperatures reached and heating times involved, and methods, with the grilling applied to the mackerel in Figure 17.5 likely giving rise, because of the nature of radiant heat applied, to a very different result that would have arisen if the fish had been baked, poached, or fried. The rate of cooking applied to the protein-rich meal examples shown is also a critical influence on the textures that arise, with the slow cooking applied to the ribs in Figure 17.9 being a key step in breaking down collagen to give a soft, desirable eating experience.

Protein-related changes that are due to cooking are also sensitive to other food components present and factors such as the acidity of the environment and the presence of salt. Another protein-rich food material which shows the mutably transformative power of heat-induced change is eggs, and the solid structures resulting from poaching in Figures 17.3 and Figure 17.4 are very different from those that might have resulted had said eggs been fried, boiled, or scrambled, because of manipulations of the preceding factors that have been practiced long before protein chemists took egg proteins apart to unravel the way in which their proteins unravel, before later scrambling them to give new forms, in both literal and molecular senses.

Similarly, as mentioned earlier in this chapter, we manipulate the texture, flavor, and cooking properties of meat when we change its internal chemistry through infusion of specific chemical agents to change the nature of the fluid bathing the muscle fibers therein, and thereby the exact unfolding and reactions the proteins will undergo when we cook them. This is the basis of the marination that led to the "jerk" in the pork in Figures 17.3 and 17.9, as jerk marinades include several forms of acid, such as lemon juice and vinegar, that will reduce the pH of the muscle and thereby alter the protein chemistry on cooking. Marinades might also include alcohol, which will also affect protein structure and meat texture, and fruit and spice components (such as ginger) might include enzymes that tenderize the meat by literally prechewing it before you get to chew it. Finally, the impact of marinating on overall meat texture is determined by diffusion, a purely physical process, as each component, perhaps at different rates, gradually makes it way from the surface into the interior of the meat, at a rate determined by, among other factors, temperature and the exact size and shape (especially thickness) of the piece of meat in question.

Throughout this book, I have shown how food preservation is a very ancient art, and some of the most traditional of preservation processes exploit some of the most potent chemical strategies of all. For example, smoking, whether applied to the salmon in Figure 17.4 or the strawberries in Figure 17.7, adds a whole world of new flavors and aromas, while making the material being smoked far more stable and microbiologically safe. Likewise, pickling the fennel and apple for the dish in Figure 17.5 may be a very traditional approach but the impact of the acid involved on the structure and texture of the vegetables involved is key to the character of the resulting dish.

In addition, traditional Asian food products such as the dashi shown in Figure 17.5 rely significantly for their particular tastes on our taste buds that are attuned to the flavor of umami, a flavor that was only officially recognized in the 1980s as being sufficiently core to our sensory profiling of food as to warrant inclusion in the exclusive club of primary tastes, which had to that date consisted of sweet, sour, salty, and bitter. One of the key sources of natural or added umami flavor in food is the sodium salt (monosodium glutamate) of the amino acid glutamic acid (proteins again!), while the compound in the fish likely giving it this savory kick is probably inosinate, a metabolic by-product of the cellular activities key to once keeping that fish alive.[4]

It has also been seen how sometimes problems with food quality arise not with bacteria or other usual microbiological suspects, but with enzymes naturally present that, when operating under conditions arising as a consequence of harvesting, chopping, or otherwise changing the natural starting material, cause deteriorative changes in the food quality. It is for this reason that 21st-century technologies such as high-pressure processing (although first reported as being possibly useful for food in the 1890s) might be used to ensure that the avocado in Figure 17.4 retained its green, fresh appearance without naturally present enzymes causing a progressive deterioration to an unappetizing brown mush.

Finally, while many of the examples discussed rely on the properties of proteins for their characteristics, in Figures 7.1, 17.6, and 17.8 we see products the construction of which relies almost entirely on the chemistry and physical properties of lipids and carbohydrates.

In Figure 17.1, the ravioli is made by mixing wheat flour or the protein-rich, relatively low-starch milled wheat fraction called semolina with water, to give a strong gluten-based matrix that is tough enough to adopt a range of shapes created by extruding or shaping the dough by using suitable tools. The pasta is then dried carefully to avoid deformation of its shape that might occur if it dried unevenly or too quickly, and later cooked in boiling water, which it absorbs, leading to swelling and

---

[4] It has been reported that human breast milk contains higher levels of umami-flavored metabolites than cow's milk, and so we may be exposed to such flavors at a very young age, which may suggest a possible alternative phonetic origin for the term itself.

softening as a result. During the cooking, the starch granules gelatinize and break open, dispersing and becoming partly soluble, while the protein becomes insoluble and coagulates, entrapping the starch and water in the expanded matrix. Getting the texture just right by judging the timing right is something every pasta cook knows instinctively how to do, without in most cases knowing that they are actually balancing competing reactions and priorities of the starch and proteins, and that allowing the balance of these to slip out of whack will result in problems like sticky, overdry, or soggy pasta.

Moving from carbohydrates to oils, the ganache shown in Figure 17.8 is an example of an emulsion, where the generally rather incompatible materials of oil and water have been induced to form a stable mixture, by the presence of molecules that can sit at the interface between globules or fat or oil and intercede on their behalf with the surrounding watery fluid, keeping the globules separate and preventing them from coalescing into a separated fat layer that floats to the surface. The ganache is made from chocolate and cream, with some butter, salt, and sugar added, and the water phase comes from the cream while the chocolate, cream, and butter supply the fat—heating it in the pan melts and mixes the fats, while the mixing disperses this fatty amalgam into droplets that are coated by natural emulsifiers called phospholipids from the cream. The rate at which the emulsion will separate is dependent on the viscosity of the liquid, as the fat droplets will move much more slowly through thicker fluids. In the case of the ganache, cocoa particles from the chocolate are also dispersed in the watery phase, and these both thicken that phase directly and bind water, rendering it less fluid. In the dispersed system formalism notation referred to earlier, this system would be described like this:

$$(O + S)/W.$$

All of these complex physical manipulations, taking components of the ingredients and rearranging them into a new multiphase structure, the behavior of which is entirely dependent on the physics of movement of large particles, collectively result in a thick and stable ganache that can then be used in a wide range of confectionary creations.

In the case of Figure 17.6, white and dark chocolate were tempered and poured into a mold the shape of an egg, and allowed to set. Tempering is a process related, as the name might suggest, to the control of temperature, and can be applied to materials from metal to eggs. The key principle is to bring a material to a temperature where inconsistencies in texture or behavior of different components present are evened out, and a more homogeneous final material (of even temper, as it were) results. Chocolate can contain a number (six) of different types (sizes and shapes) of crystals of cocoa butter (fat), some of which give good textures to chocolate, but some, if they dominate, can give chocolate that is too soft, too hard (and crumbly), or too easy to melt. In tempering, we first melt all the crystals present, and we then gradually decrease the temperature with careful stirring to induce only the most preferred forms to reappear, sometimes warming back up slightly to tailor the process very delicately, giving (hopefully) a smooth and nice-looking chocolate.

The "eggshell" was then filled with a coconut mousse (where air has been incorporated to give a light fluffy texture, just as happens when a real egg is transformed
into a meringue) to resemble the white of the egg, while the yellow part is mimicked
by a passion fruit sauce that has been given a "fluid gel" consistency by adding a
polysaccharide called gellan gum. Gellan gum is produced by a bacterium called
Sphingomonas elodea, which helpfully excretes it from the cell when produced, so
that it can be grown in fermentation tanks and the gum recovered quite readily for
purification into an ingredient for, among other things, fluid fruit gels that mas-
querade as egg yolks. This is also a useful reminder that not all points of intersection of the microbial world with food result in spoilage or illness, and this is just one
of of many examples in this book that rely in one way or another on harnessing the
actions of beneficial microorganisms, from wine to beer.

The reason gellan gum is an interesting kitchen ingredient is that it gives very
nice gels that will be thick but yet flow (fluid gels) at low levels of addition, without
needing to be excessively heated (and so delicate flavors are not sacrificed in the process of making the gel). The idea of a fluid gel sounds paradoxical, as we think of
a gel as something reasonably solid(ish) and nonflowing while the fluid part of the
term sounds, well, fluid. In the case of gellan gum, once it makes a gel (by cooling
a hot solution of the gum in the presence of particular minerals) it is indeed reasonably solid but, like xanthan gum (another example of polysaccharide made by
bacteria that is now increasingly found as a powder in kitchens) when a gellan gum
gel is subjected to a small force (like pouring it), it becomes a lot less viscous and
flows. This happens because the gel essentially divides under this force into smaller
units of gelled particles, which can flow past each other in a thick (indeed eggy
runny) mass.

So, polysaccharides can be induced to act like proteins in the right applications,
and the principle of fluid gels can be found in the kitchen in products like ketchup
(which thins significantly as more force is applied to trying to get it to flow) and
chocolate milk (where, at rest, polysaccharides keep the chocolaty cocoa particles
suspended and don't allow them to sink to the bottom of the container). The secret of the difference between a fluid gel and a liquid that is simply very thick is the
relationship with force. Some thickeners will give a liquid that has the same thick
consistency under any conditions it is likely to encounter, but a fluid gel will, when
force is applied, become a lot less gel and a lot more fluid.

Overall, these examples have hopefully shown that the claim in the Introduction
of this book that a kitchen is a highly scientific environment, in which physics,
chemistry, and microbiology are routinely exploited, and scientifically elaborate
reactions casually induced, through a range of processes that have more in common
with those used in large-scale food production than perhaps expected, in the preparation of meals. Preparation of every dish is an experiment, in some cases more
highly controlled than others, which results in the metamorphosis of molecules
into meals.

{ 18 }

Final Thoughts

It seems appropriate to finish this book with the equivalent of a dessert or aperitif, to send the reader off with a sense of satisfaction, satiation, and hopefully pleasure.

I thought about polishing my crystal ball and trying to project into what food might look like in the future but, as the Nobel Prize–winning Danish physicist Niels Bohr once said, prediction is very difficult, especially when it is about the future. Futuristic predictions are of course notoriously unreliable, as can be seen by the fact that we should all surely have our personalized jet-packs by now.

Interestingly, one theme that may have come through in this book is that the future of food, at least for the next few decades, is, to adapt a quote by the writer William Gibson, probably here already, but just not equally distributed. The progress of food science has happened sporadically and unevenly, as when Bert Hite showed that high pressures could preserve food a century before anyone figured out how to make that work in a practical sense, and when NASA was introducing innovations in food safety and packaging for space travel that years later have become common practice in our restaurant kitchens and on our supermarket shelves.

The story of food science in the last century has been about taking all that we knew about the art, provenance, and processing of food in the pre-scientific era and underpinning anecdote with fact and understanding. I think that this great era of scientific study of food has answered the main questions, such that we understand broadly *why* most of the things we have observed since mankind emerged and started to eat things happen, and moreover how to control these to our greatest advantage. Many scientific phenomena relating to food are well described, in textbooks, websites, and a huge body of scientific papers, while of course leaving plenty of interesting questions and challenges for future generations of food scientists to explore.

The modern food production environment, whether small or large, for example, making meat, cheese, or beer, is today a much more scientific (explicitly or implicitly) environment, with degrees of safety, consistency, and efficiency unthinkable when similar products were made for much of their history, thanks to food science.

The combination of the key twin disciplines of food science and nutrition has helped to ensure that understanding of what exactly food does in our bodies, for good or ill, is matched by practices, formulations, and processes that maximize the nutritional benefit of our food. Nutrition sets the parameters for food science and keeps the importance of food as the most significant chance to change our bodies and lives three or more times a day (if we are lucky) at the forefront.

There is no doubt that the range of available food products will diversify in the future, probably less in terms of the processes available and raw materials used than in the way in which these are combined. It is inevitable that we will in the future encounter products that are categorized and sub-categorized, with different flavor, nutritional, and even textural properties, for every lifestyle and life stage, from birth to old age, even more than we do today. In such a future, students might eat quite differently to athletes[1] to those seeking whatever lifestyle might be of interest to them at a particular time. When we are sick, we will have food products tailored to our specific needs to supplement medicine to a greater extent than ever before, whether having textures engineered for those who have impaired chewing and swallowing or being designed to provide the protein required to fight muscle wastage during prolonged illness. We might even soon walk into a supermarket or step up to a vending machine and specify to a person or touchpad a set of requirements for the snack or meal we desire at that precise time, and have it produced by a 3D printer there and then.

We will certainly be better informed, as the vast knowledge of food becomes available to us in convenient form, as we use phones or other smart devices to scan food packages or even meals in restaurants, to augment that reality with details about the nutrition, origin, and perhaps safety of the product in question.

These are all developments that are happening already, and will just continue, probably at an accelerated pace.

Another key development discussed in this book has been the dissolution of the barriers between the worlds and mind-sets of chefs and food scientists and, whether we use terms like "molecular gastronomy" in 10 years or not, that new relationship cannot fail to generate new sources of innovation and practice in kitchens around the world.

Nonetheless, as I type the words "around the world," I am acutely conscious that much of what I have described in the preceding paragraphs is really of relevance to a minority of the world's population, who have comfortable, affluent lifestyles, and that for huge swathes of humanity basic access to clean, safe food and water is not a lifestyle choice but a constant struggle. The responsibility of food science to provide developments for those for whom molecular gastronomy and 3D printers are as irrelevant as space travel cannot be understated or ignored.

---

[1] Young professional sportspeople could consume pro teen proteins, for example.

So, as I come to the end of this stream of consciousness, and indeed this book, I hope you the reader will have changed your perspective on food, as was my stated objective at the outset, appreciate to a greater extent the contribution of science to almost every aspect of food, and realize that there genuinely is more to food than meets the eye.

On the Analysis of Cheese

How do we see and study proteins in food?

An interesting question is how we "see" proteins in food, to study them. In the case of cheese, which I will use here to explain the principles of the key methods concerned, there is a whole array of techniques we can apply. Most literally, we can see them by using a powerful electron microscope, which reveals the structure of cheese to be composed of chains of casein micelles assembled into a three-dimensional network, like tiny strings of pearls, created by the action of rennet (as in the images discussed in Chapter 3). Within this scaffold can be seen large fat globules, trapped inside as the liquid serum surrounding them suddenly solidified into a set of cages from which they could not escape, and clusters of bacterial cells, mostly from the starter culture added not long after the rennet, but also perhaps arising from the cheese vat or even the air surrounding it.[1]

To see the proteins less literally, but more specifically in terms of their profile and changes during ripening, a common technique used is called electrophoresis. In this method, a thin sheet of a fairly nasty material called polyacrylamide is clamped vertically in a chamber through which an electric field can be passed, such that each end of the gel sheet has an electric charge, but of opposite natures (negative and positive, like in a battery, at top and bottom of this vertical sheet, respectively). When a mixture of proteins is loaded onto the top of the gel, it is mixed with solutions that make sure the protein molecules are all highly charged (negatively). In this circumstance, they will feel a strong attraction toward the far (positively charged) end of the gel, which is due to the fact that opposite electric charges

---

[1] These "nonstarter lactic bacteria" and their significance, was discussed in Chapter 7. In cheese science, to refer to something as a nonstarter is not a dismissive term, and indeed as these grow mostly after the added starters have died off, "continuers and enders" might be even more descriptive of their role.

attract. In parallel, they are kicked away from their start line because of its negative charge, which repels the proteins because of their similar charge.

So, a race to the finish line is started when the electric field is turned on, and all the proteins take off. However, the racetrack is not a smooth track, but rather an assault course, with the polyacrylamide gel forming a dense entangled network through which proteins need to fight their way, like a person forcing his or her way through a thorny bush. Smaller proteins can force their way more quickly and move more rapidly, while larger protein molecules lumber more slowly. By the time the first proteins (the smallest) reach the far end of the gel, they have left their heavier counterparts behind at points that correlate directly with their size, and the heaviest proteins will have moved the least from their starting line.

So, proteins are separated on the gel in a manner that reflects their size, but can we now "see" them? The polyacrylamide gels are like translucent jelly, and the proteins cannot be seen yet. To make them visible, we need to develop the gel, like old camera film, and this is achieved by immersing the gel in a solution of a suitable stain, which binds to the proteins and lends them a highly visible color, typically a deep purplish-blue. Now we can see the proteins, and typically a mixed sample of proteins (of which several can be separated side by side on a gel several inches long and wide) forms a very distinctive bar-code-like pattern, of short "bands," each of which corresponds to the separated molecules of one protein that has moved to that point on the gel while the electric field exerted its inexorable forces, on the basis of its size.[2]

We can then record this profile electronically or by photography[3] and use computer software to measure the position and intensity of the bands, and so the level of proteins present.

If we analyze a sample of fresh cheese using this technique, we will see exactly such a protein bar code, with one line in the bar code for each of the caseins, which have been collectively concentrated into the cheese curd. In fresh cheese curd, we will see four major caseins and a few of their initial large chunks released by the enzymes present in the milk or added rennet, and so the bar code is pretty simple.

Let the cheese sit in its ripening room, though, and the magic starts to happen, so the code becomes more complex. Load a sample of cheese taken after a few weeks or months of ripening, and the pattern will have changed in two ways. First, the bands that were there to begin with will have become fainter and will progressively shrink away, as they are digested by the huge array of enzymes present in the cheese, which are essentially chewing away the caseins, long before consumers get

---

[2] In milk and cheese, the caseins have annoyingly similar sizes, and don't tend to pull apart very well, leading to a clump of bands closely together on the gel that are hard to differentiate into the different proteins present. Thus, for these applications food scientists usually use a slightly modified technique that separates proteins based on a combination of their size and charge, which is a better way to see the different types of casein clearly apart from each other.

[3] This just involves pointing the camera and instructing "say cheese."

their chance. This can be seen in the top part of the schematic diagram of Figure A.1 in this Appendix.

In parallel, the initially simple pattern becomes more complex, as new bands appear, the products of this initial enzymatic gorging. As one casein molecule is chopped, for example, in half, two new products appear, comprising the amino acid chains on either side of the point of breakage, and each of these will have a different behavior in the electrophoretic separation to each other, and their parent protein. So, the parent band becomes fainter with each of the molecules that gets chopped up, and two new bands have appeared.

As time goes on, though, some of these bands may become progressively fainter, as other enzymes break down the half-protein fragments, and so these too may pass away, and fade. In theory, given enough time,[4] an electrophoretic gel of cheese might become quite blank when all the proteins have disappeared.

How do we see smaller fragments from proteins in food?

Where do the proteins and their identifying bands go? The staining we use to "develop" the gel and visualize the protein bands fails when the protein fragments become too small to absorb sufficient stain to announce their presence, and small peptides and amino acids will essentially become invisible. This also happens because the smaller fragments can run unhindered right through the gel like a bad case of something you don't want to catch from your food.

To "see" these, we need a different technique that detects these short and small products, identifies their presence, and measures their level. Again, the key to such a method must involve some principle whereby a complex mixture can be separated on the basis of some property that differs in a predictable and understandable manner between the elements of the mixture, combined with some way to visualize the components thus separated.

In the case of analysis of small peptides in food, this involves what is called chromatography.[5] To understand this, picture a tube filled with (extremely) small beads, which have been chemically designed such that they attract peptides to stick to them, but where the property by which such attraction occurs will differ between peptides, for example, because of their different make-ups in terms of amino acids. One property might be charge, whereby the peptides have different levels of

---

[4] For example, through a fortuitous experiment involving a sample of cheese found lurking (perhaps having evolved into something quite new, and possibly alive and furry) and forgotten for many years at the back of a cupboard or fridge. Or perhaps we need to search for bog cheese!

[5] As the name suggests, chromatography was initially developed to separate colors, and in its simplest form can be seen when ink is spilled on some paper or other material, and water is introduced. The colors will flow with the water, and the various constituent colors that make up the compound color of the ink can be separated if they move at different speeds. Occasionally, this separation is conducted using a piece of equipment called a washing machine.

electrical charge and the beads have the opposite charge, and so the peptides adhere to them by electrical attraction between oppositely charged molecules.

To make separation happen, we pump a liquid containing the peptides through the tube (in this method called a column). The liquid that initially comes out the far end is depleted of the peptides, which have stayed behind huddling on to their newfound mates that they have found (electrically speaking) irresistibly attractive.

Now we need to change the nature of the liquid flowing through the column past the beads, to make the conditions in the liquid progressively more attractive to the peptides than the bead surface, or to introduce something that fights for the binding spots on the beads with the peptides. At first hint of these more conducive conditions, the most weakly bound peptides will fall off easily and flow out through the column. As we progressively increase the key attractive property of the flowing liquid (called the eluent) more strongly bound peptides will gradually be enticed to go with the flow, as it were. An imperfect analogy that comes to mind is dirt on a garden path that is being targeted by a garden hose; at low velocity; the most loosely adhering dirt will be blown off, while getting the more tacky stuff to shift will take stronger and stronger flows.

Moving back to the column, the eluent, as it comes out the far end, thus changes over time and contains at different times different peptides, which have separated on the basis of the property exploited in the first case by the beads to get them to stick, and then in the second by making the liquid progressively more favorable an environment than the beads.

Now we need to "see" and measure these compounds. In most cases, this is done by shining a beam of light through the liquid in a transparent cell, with the presence of proteins and peptides being spotted by an increase in the way in which that light is absorbed (as bonds within the molecules interfere with the passage of the light).

A typical separation might take 30 minutes, during which the composition of the liquid flowing through the column has been gradually changed in a controlled manner.[6]

In the case of cheese, what will be run through the column is an extract of the cheese, perhaps made as simply as by mashing the cheese in water, in which the intact caseins and their large products are insoluble, but in which the peptides will generally be more soluble. The output of the separation of this sample on a chromatography column is a graph of the output of this detector versus time. Each peptide will show up on this plot as a peak or spike in the detector signal.

So, on analysis of fresh cheese, in which very little proteolysis has taken place, very few peaks will be seen, as there will be few peptides to extract from the cheese. Give it time, however, and the picture becomes ever more complex, and the initial flat baseline from the detector grows first a few spiky peaks, and then an

---

[6] This can be achieved by having two different solutions that can flow through the column, which represent opposite extremes of the condition that attracts the bound peptides to be released. This might be as simple as low and high levels of salt, and gradually changing the proportions of the low- and high-salt solution flowing though the column is called a gradient separation.

ever-more-complex forest of these markers of the presence of a peptide, each one telling a tale of its origin (the protein whence it came, and the enzyme that liberated it), and its contribution to the flavor and character of the cheese.

Thus, as the electrophoresis gel gets first busier but then progressively emptier, so in parallel the complexity of what can no longer be seen there is captured on our column. This can be clearly seen in the lower schematic diagram of Figure A.1. of this Appendix.

In practice, the columns on which these separations are carried out are thin (millimeters in diameter) and a few centimeters long, and the beads packed therein are microscopically small, with better separation coming with smaller beads (binding more peptides and releasing them more selectively). To force liquid through such a column is not easy, and requires high pressures, and so the technique is called high-pressure (or sometimes high-performance) liquid chromatography, or HPLC, one of the key workhorse techniques of a food analysis lab. The principle by which the peptides are induced first to bind, and then inveigled to escape, can be based on the molecules' fear of water in one commonly used method, whereby the constituent proteins are separated on the basis of their hydrophobicity (or relative affinity for water). Such separation is called reversed-phase HPLC, or RP-HPLC, and the resulting separation of peaks shows the analyst that the early eluting peaks are quite water-friendly, but the later ones (pulled tenaciously off by increasing the level of organic solvent passing though) are more water-suspicious, and the profile is essentially a profile in hydrophobicity. As caseins and the peptides generated from them in cheese are relatively hydrophobic to begin with (hence the reason why the water was able to leave them out of the extract to begin with), this works pretty well for cheese peptide separation.

So, after electrophoresis or HPLC, we have a picture, in one case a sort of bar code and in the other a forest of peaks, and we can then compare one type of cheese to another, and see how a piquant Blue differs from a simpler Quarg, or we could compare samples of cheese at different stages of ripening and note how they have changed, as age exerts its biology on the initially simple set of proteins we have snatched from the milk and squished into this cheesy mass.

But, what we cannot say is what these bands or peaks are or where they come from. To help with this, we could in one approach run standard purified proteins alongside our sample being analyzed, compare the migration behavior, and on this basis make informed judgments of what is what based on similar separation behavior.

However, many of the bands and peaks will remain unidentified as we won't have standard comparators for the possibly hundreds of fragments that will be released during ripening, and again that differ from cheese type to cheese type and over time too.

To understand how we perform such identification, let's consider someone bringing a vase into an antique shop, which a quick informed glance will indicate is old, possibly very old, and possibly very valuable. Now imagine if the expert who asked the owner's view on the provenance of this artifact reached behind the counter and brought out a large mallet and proceeded to smash the vase into tiny pieces.

Once these pieces are sufficiently small, the expert then sifts through the fragments, and from these eventually and with unquestionable certainty pronounces to the likely dismayed owner of this former fine vase the exact origin and identity of the object.

This is a bit like the best means we have to identify proteins.

We use a machine called a mass spectrometer, which can measure the exact mass[7] of molecules to incredibly precise degrees of exactitude. Each protein or fragment of a protein is, as we have discussed, a particular and unique sequence of amino acids strung together. The mass of the peptide or protein is then the sum of the mass of the amino acids it is made of, added up together, and, if every such protein or peptide is defined by its unique combination of amino acids, so each has a different possible mass.

In the mass spectrometer, the molecules whose masses are to be determined are again set off on a racetrack, like the gel in electrophoresis or the column of a HPLC system, and again separated by the rate at which they progress down this track. In most of these machines, essentially a laser blasts the molecules into a highly excited and mobile state,[8] and they are then carried through a chamber long enough to separate large from small, until the exiting molecules again pass a highly sensitive detector, which records their passing and converts this to an atom-precise determination of their mass.

For this to work well, large proteins aren't the best things, being hard to excite sufficiently and get moving, and so we bring out the metaphorical mallet. The proteins are treated with an enzyme (molecular scissors alert once again) that cuts them up in a highly nonrandom and predictable way, into a bunch of fragments (like the shards formerly known as vase), which will migrate just fine in the mass spectrometer. Then, each protein gives a series of masses, and each protein will give only one combination of masses, and so the results of the detector's determinations can be plugged into a computer database, containing the "fingerprints" of huge numbers of proteins for which this has been worked out and the original identity ascertained.[9]

So, to get to this bit, the protein band on a gel can be recovered (sometimes using something as simple as a blade), the gel dissolved out from behind it, the

---

[7] The mass of a molecule is the mass of the atoms it comprises. Needless to say, atoms and molecules are rather small, so we can't use regular units like grams or ounces, but rather we use a scale that effectively compares the mass with that of the simplest atom possible, a hydrogen atom, which has a weight of one. On this scale, amino acids might have a weight of a few dozen, peptides a few hundred, and proteins in the thousands. The units used to "weigh" molecules are called daltons, after the famous biochemist who proposed the scale and, when we count in the thousands for proteins, we refer to kilodaltons. I guess it is lucky the scale was proposed by someone whose name doesn't sound too odd in this usage—try inserting your own surname after the suffix "kilo" and see how it sounds.

[8] To be fair, I think any of us would become rather excited and move rapidly if blasted by a laser, and maybe would be a little jealous that molecules have entered the world of Star Wars and other science fiction ahead of the rest of us.

[9] This always reminds me about the old joke about the hardest jigsaw in the world being a bag of crisps or chips from which one is told to assemble a potato.

enzyme added, and the mass determined as indicated. Now each band has a mass, a name, and an origin, and even perhaps a good guess as to why it is there and what it is doing.

For the HPLC profiles, the work has been half done, as what are separated are peptides or fragments already, and so the mass spectrometer can work immediately on the fragments coming off the column if hooked up together.

It is ironic that we can take a wonderfully character-filled cheese, perhaps made in the most traditional of circumstances, and for which a connoisseur can describe the exact aroma, flavor, and texture, and then deconstruct it in a set of machines that look incredibly high-tech. These marvels in stainless-steel casings enwrapping unimaginably complex innards cost tens or hundreds of thousands of dollars, and linked computers both tell them what to do and crunch through their outputs, to translate the personality of a cheese into a list of numbers and esoterically coded protein identities. Again, does this ruin the magic or tarnish the wonder? Not at all in my opinion, as it shows the complexity of what nature and the cheesemaker have worked together to create, while being a testament to the craft that worked out how to unpick the molecular secrets of cheese.

{ FURTHER READING SUGGESTIONS }

When I started writing this book, I immediately jumped into areas where I was comfortable after two decades of teaching and research, mainly those concerning dairy products and processing, but to present the whole picture of food I needed to read widely and learn much. This included a huge introduction to the science of the kitchen, a particularly notable journey for someone once described by one of Ireland's leading chefs as the food scientist who can't even cook an egg. This learning journey over a year or so was helped enormously by several individuals thanked in my acknowledgments, but was also greatly assisted by some great books, which are listed in following, along with some comments.

Bamforth, C. *Beer*. Oxford University Press, New York. 2009. [A pretty definitive overview of the eponymous beverage, covering almost every imaginable aspect]

Barham, P. *The Science of Cooking*. Springer-Verlag, Berlin, Heidelberg. 2001. [A very nice overview of a lot of aspects of the science of what happens when we cook, with great examples of the science behind a wide range of recipes]

Davidson, A. *The Oxford Companion to Food*. Oxford University Press, New York. 2014. [An excellent and enormous reference work, combining historical, cultural, and scientific perspectives on a vast array of food topics]

Field, S. Q. *Culinary Reactions—The Everyday Chemistry of Cooking*. Chicago, Chicago Review Press. 2015. [Full of very nice and accessible facts and anecdotes about kitchen chemistry]

Handel, A. P., Schloss, A., and Joachim, D. *The Science of Good Food*. Robert Rose Inc., Ontario, Canada. 2008. [A really useful alphabetical encyclopedia of food]

Jopson, M. *The Science of Food. An Exploration of What We Eat and How We Cook*. Michael O'Mara Books Limited, London. 2007. [A concise exploration of some very nice examples of how science underpins food]

McGee, H. *On Food and Cooking*. Simon & Schuster, New York. 1997. [The classic and probably definitive work on the science of the raw materials and transformations of the kitchen]

Myhrvold, N., Young, C., and Bilet, M. *Modernist Cuisine: The Art and Science of Cooking*. The Cooking Lab, Bellevue, Washington. 2015. [A giant piece of work, in every sense, weighing several kilos and coming in five large-sized hardbound volumes, in a case, with an accompanying workbook, and including incredible photography and a surprising amount of detailed science, along with enough recipes and ideas to keep the scientifically minded chef busy for years. Worth the enormous price, and a new equally voluminous companion set on bread has just been published.]

Potter, J. *Cooking for Geeks*. O'Reilly Media, Sebastopol, Canada. 2015. [Contains a lot of interesting scientific discussions of food, recipes, and concepts, and contributions from a wide range of chefs and others interested in food]

Shephard, S. *Pickled, Potted and Canned—How the Preservation of Food Changed Civilisation.* Headline Book Publishing, London. 2001. [A very readable account of the history of food processing]

Spence, C. *Gastrophysics—The New Science of Eating.* Penguin Viking, New York. 2017. [A very entertaining and thought-provoking overview of the role of psychology and related factors on the experience of eating, in contexts from food packaging to restaurants]

This, H. *Molecular Gastronomy—Exploring the Science of Flavour.* Columbia University Press, New York. 2006. [A key reference outlining the mission statement of the discipline of molecular gastronomy, with short chapters on a variety of topics, from historical aspects of food through flavor and ingredient science to visions for the future]

Tunick, M. *The Science of Cheese.* Oxford University Press, New York. 2014. [A good overall introduction to the science of cheese]

Vega, C., Ubbink, J., and Van der Linden, E. (eds.) *The Kitchen as Laboratory. Reflections on the Science of Food and Cooking.* Columbia University Press, New York. 2013. [A great collection of short articles on different topics by many experienced food scientists, often taking an unusual product and explaining the science behind it]

Wolke, R. L. *What Einstein Told His Chef.* Norton, New York. 2002. [A very enjoyable book that approaches a broad sweep of topics through a series of questions and propositions to be tackled, and using science, anecdote, and recipes to do so]

{ INDEX }